Water:
The International Crisis

Robin Clarke

The MIT Press
Cambridge, Massachusetts

First MIT Press edition 1993
© 1991 Robin Clarke

This book was printed and bound in the United States of America.

This edition published by arrangement with Earthscan Publications Limited, 120 Pentonville Road, London N1 9JN, England.

Library of Congress Cataloging-in-Publication Data

Clarke, Robin.
  Water : the international crisis / Robin Clarke. — 1st MIT Press ed.
      p.   cm.
  Includes bibliographical references and index.
  ISBN 0-262-03208-2 (hc). — ISBN 0-262-53116-X (pbk.)
  1. Water-supply. 2. Water conservation. I. Title.
TD345.C59 1993
333.91—dc20   92-36487
                                                                    CIP

Contents

Acknowledgements

This book owes its origins to a meeting held in February 1989 in Vadstena, Sweden, on the way water scarcity in semi-arid regions increases human vulnerability. Convened by the Swedish Red Cross and the Department of Water and Environmental Studies of the University of Linköping, the meeting brought together experts from many different fields to debate the alarming and intensifying effects of water scarcity in the world's drylands.

The meeting identified a number of ways to attack problems induced by water scarcity. Among them was the need to create awareness among donors and policy makers of the extent of the water-scarcity problem, and of the range of possible solutions to it.

This book, which has been funded by the Swedish Red Cross, results directly from the Vadstena meeting. A summary of the meeting is included as Appendix I.

My special thanks go to Professor Malin Falkenmark, of Stockholm's Natural Science Research Council whose work inspired this book. None of its defects, however, are her responsibility.

The International Red Cross and Red Crescent Movement with its National Societies, International Committee of Red Cross and League of Red Cross and Red Crescent Societies in Geneva has established a worldwide network to alleviate human suffering. However, during recent years the work has also been focused on how to prevent disasters. It is within this context that the Swedish Red Cross wanted to support Earthscan in giving more emphasis to the fundamental role of water in making particularly semi-arid regions less vulnerable and prone to disasters.

Preface

For six millennia, the human race has been involved in a battle to control water resources. Impressive works and facilities have been constructed to harness these resources to the benefit of humankind and society. But just as water is a resource with a certain evasive character deployed in its eternal move through the hydrological cycle, its importance for society and for the Earth system seems to have slipped out of the imagination of policy makers and development architects alike. Every project and almost any process in society and in nature need water, but few mention it. Invariably, water is taken for granted. A kind of water blindness has become common in discussions about development projects and even in key policy documents about our future.

As a contrast to a predominant simplified notion of water stands the complex and multifunctional role that water plays. Water is the lifegiver *par excellence*. Besides coal, water is the most fundamental substance making life possible on our planet. While most resources have substitutes, water fulfils a number of functions where there is no substitute. In any conceivable scale, from the global circulation system down to the individual cell, water is present. It stores, redistributes and releases about thirty per cent of the total amount of solar energy that hits the Earth. But it is also a fragile resource, and for people in large parts of the Third World, in particular, it is increasingly a scarce resource. Apart from the arduous daily task of securing even the small amounts required for survival, its benign functions are sometimes overshadowed by deleterious effects. Polluted water and water quality degradation are estimated to be the cause of up to eighty per cent of disease.

Even if water is a renewable resource, it is at the same time finite. Its availability is largely dictated by climate. Low precipitation in combination with a high evaporative demand by the atmosphere mean that the amount of water that remains and can be put to use is small. Moreover, it fluctuates from season to season and between years. It would be a serious mistake, however, to interpret water scarcity only in hydrological terms. Seasonal water scarcities are being reported from areas that have quite high rainfall regimes. As noted in this book, the area around Cheerapunji is actually the wettest "desert" in the world. In spite of an annual rainfall of more than nine metres (9000 mm) – a world record – the landscape is desiccated only a couple of months after the rains. This astonishing example highlights the dramatic and disastrous consequences of a resource management where the interrelationships between water, land and vegetation are not duly considered.

During the last decades, much progress in the efforts to improve living conditions all over the world has been achieved through technological solutions. Total water use in the world has quadrupled during the last fifty years. At present, about forty-four per cent of the world's water resources in terms of reliable run-off are being made accessible through the construction of dams, reservoirs and conveyance structures. This is a high figure, and even the most optimistic calculations do not give much hope for any substantial increase in the fraction of water that can be withdrawn. For the billion or so people living in the semi-arid regions of the world, the vagaries of nature have to some extent been halted and controlled. But due to economic costs and to their social side effects, the heyday of construction of large dams and transfer systems is over. Currently, and for the future, the livelihood conditions for the burgeoning populations can only marginally be improved through such measures. Storage of water is still a must, but it is increasingly important to make sure that the considerable losses in terms of evaporation and seepage that accrue from reservoirs and conveyance systems are halted.

As emphasized in the book, there are still untapped water resources and also options to make better use of scarce water resources. Various means of water harvesting provide a means to increase the amount of rainwater that can be utilized. For the amount of water that is available, the challenge is to allocate it to the best possible use. By choosing the "right" crops or allocating it to the most productive sector, the benefits derived from water use can be substantially enhanced. Efficiency in water use is generally low in projects built during the last few decades. In addition, the productivity of the water that is accessible is low. For the future, efficiency in water handling and its allocation for productive and optimum use are crucial.

Taking the mounting water scarcity seriously requires an increased awareness among policy makers and development planners, at all levels in society, of the significant role of water for development and in nature. New approaches are needed for the proper management and use of water resources. Instead of asking how much water we need and where to get it, we should ask how much water there is and how we can best benefit from it. Much of the world is in for a bad time. But in a period of so many opportunities it is not worthy of the international community or any group or individual to succumb or surrender to the Malthusian thesis.

Robin Clarke's book provides a holistic view of one of the essential resources serving humankind and making life possible and worth living, but which has been seriously neglected, mistreated and often poorly managed. As the author states, it is not the job of this book to invent a political programme that will bring water to the forefront of the world's action agenda. Its job is to illustrate the urgent need to do so. It is, indeed, not only an urgent need. When attended to, it may yield significant payoffs in the efforts to create a decent livelihood for the population of the world today and tomorrow.

Addressing water blindness in general, the book has a relevance for all countries and climates. However, it should be of particular concern for people living in semi-arid areas, which are prone to disasters. Within the international community governments and non-governmental organizations (NGOs) alike should be able to use the book in the planning process to create more stable living conditions and avoid droughts and other tragedies. Prevention is always better than cure.

Thus we hope that this book, within the framework of the United Nations Conference on Environment and Development (UNCED) in June 1992, as well as within the International Red Cross Movement – the two organisms representing all governments and NGOs – will serve the purpose of making everybody better respect our most basic precious resource on Earth: water.

Maurice Strong	**Birgitta Dahl**	**Gudrun Göransson**
Secretary General	Chairperson	Chairperson
United Nations	Swedish National	Swedish Red Cross
Conference on	Committee for	
Environment and	UNCED	
Development	Minister of the	
(UNCED)	Environment	

Chapter 1

Water Scarcity

All peoples . . . have the right to have
access to drinking water in quantities and
of a quality equal to their basic need.

UN Water Conference, La Plata, 1977

At the end of 1984, 21 African countries were suffering from what the United Nations calls "abnormal food shortages". In human terms, this meant that hunger stretched its dread hand in a belt across and down Africa, starting from the Cape Verde Islands off the west coast, moving over the Sahel to Ethiopia and the Sudan, and then extending down into Botswana, Mozambique and Lesotho.

In this area of more than 14 million square kilometres – one and a half times the size of the whole of Europe – some 200 million people, 40 per cent of Africa's total population, didn't know where their next meal was coming from. The death rate in some of the emergency camps that were set up to house the starving was more than 100 a day. By the time the crisis ended the following year, hundreds of thousands of Africans had either starved to death or died indirectly from malnutrition.

A similar event, causing millions of deaths, had swept through Africa in the early 1970s. Why should famine hit the same continent so disastrously in the space of two short decades? What caused the food crises of the 1970s and 1980s?

Drought, world recession, and civil and military conflicts all played their part. But the real truth was that Africa was short not so much of food as of water.

A shortage of water is not the same as drought. Droughts are exceptional meteorological events. Water shortage in much of Africa – and elsewhere – is not exceptional but endemic, a part of everyday life. The central thesis of this book is that many countries, and not only those in Africa, are now chronically short of water. Most of them are likely to become more so in the future.

While the symptoms of water shortage are easy to identify, its causes are not. Malin Falkenmark, from Stockholm's Natural Science Research Council, distinguishes four different causes of water scarcity:

- aridity, a permanent shortage of water caused by a dry climate;
- drought, an irregular phenomenon occurring in exceptionally dry years;
- dessication, a drying-up of the landscape, particularly the soil, resulting from activities such as deforestation and over-grazing; and
- water stress, due to increasing numbers of people relying on fixed levels of run-off.

The first two of these relate to the climate, the second two to changes that result from human activity. For the moment, it matters little which of these causes is the most important, and which results in the greatest suffering. Suffice that, jointly or separately, they are depriving millions of people of the water they need to live anything approaching a decent life.

Water shortages are also a potential problem in many developed countries. However, industrial nations can usually resort to buying their way out: through the use of expensive energy, expensive technology and expensive investments they can install the wherewithal to recycle their water, or even to desalinate sea water.

Developing countries, trapped in poverty and debt, have no such option. Those that suffer from serious water shortages are faced with a cruel dilemma: they must either limit their use to water that has not previously been used; or they must make do with used but untreated water.

There are few meaner examples of Hobson's choice. To choose to limit water use can prevent progress, reduce food production, threaten livestock production and imperil human survival. To choose to reuse untreated water, on the other hand, is an open invitation to disease.[1]

The Costs of Water Shortage

What does this mean in practice? A shortage of water can prevent almost everything from being done. Consider the following statistics: you need at least 3 litres of water to produce a tin of vegetables, 100 litres to produce one kilogram of paper, 4500 litres to produce one tonne of cement, 4.3 tonnes to manufacture one tonne of steel, 50 tonnes to manufacture a tonne of leather and no less than 2700 tonnes to make a tonne of worsted suiting.

Even more importantly, the average human – of which there are now more than 5 billion on the planet – needs to drink a litre or so of water a day to stay alive. That is, if he or she is adequately fed. Water requirements of those who are on starvation diets are dramatically higher, because food itself consists mainly of water.

But this is not the most important statistic of all. There are few places in the world where survival is threatened directly by lack of drinking water. But there are many where a lack of food apparently threatens survival. As this book will make clear, it is often not food that it is in short supply in these places, but the water with which to grow it. To grow an adequate diet for a human being for a year requires about 300 tonnes of water – nearly a tonne a day. There are many, many places where there is not nearly enough water to do this.

Water is thus the one essential requirement of all forms of food production. No water, no food. It is no coincidence that fewer people go hungry in wet countries than in dry ones .

Water is thus a limiting factor in human development, and water shortages are heavily implicated in humanity's present plight. If better means of conserving and using the water we have are not found, development will remain slow in many of the poorest areas of the world for decades, possibly centuries.

Water deficits are not restricted to developing countries, nor even to arid or semi-arid areas. The more water that is needed, the more likely is a prospective deficit. Industrial countries, which know well the value of the water they use, will go to immense lengths to avoid the deficits that might threaten to halt their development. In the Soviet Union, for example, the north of the country has plenty of water but water deficits in the south are expected to reach as much as 100 cubic kilometres a year by the end of the century. One (recently cancelled) plan to overcome the deficit involved altering the course of northward-flowing rivers towards the south, through the construction of transcontinental canals. China has already undertaken projects of this type and the United States has plans to.

Such grandiose schemes are fraught with problems. They have unpredictable ecological consequences; except in the very largest countries, they require complicated international agreements that can be enforced; and they require massive funding in a market still fighting over what should be done over repayment of previous levels of Third World debt. Furthermore, in many areas of the world, water-rich areas do not lie near enough water-poor ones to make such schemes economic. Most of the developing countries that are short of water are therefore condemned to a different, and unsavoury, solution: the use of dirty water.

Welfare and Illfare

The success of civilized society is due less to the invention of the steam engine, the spinning jenny and the other technological wonders of the industrial revolution than it is to the invention of cleanliness. Had Dr John Snow, in 1854, not been able to trace an outbreak of cholera to the Broad Street Well in London's Golden Square, we might never have emerged from Dickensian squalor. But he did, and in the process of showing how the well was being contaminated from a nearby privy used by those carrying the cholera bacteria, he revealed for the first time the link between disease and water that has always been man's greatest plague.

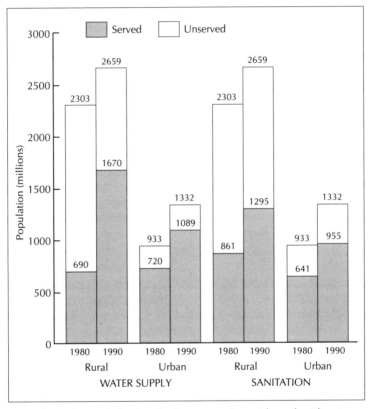

Number of people in developing countries with and without an adequate supply of drinking water and sanitation; 1980 and 1990

But the birth of sanitation was a troubled one. After Snow's discovery, the London authorities ordered that wastes be discharged into the run-off system that carried storm water to the Thames. The load of organic matter thus deposited in the river was more than it could absorb, and the stench became so intolerable that the Houses of Parliament were forced to hang burlap sacks saturated in chloride of lime in their windows so that Members could continue the business of the day. This no doubt hastened the development of modern sewage treatment, which first began to be practised in the late nineteenth century.[2]

In developing countries, however, most drinking water is contaminated and most sewage is left untreated. A 1975 survey by the World Health Organization (WHO), which covered 90 per cent of developing countries (excluding China), showed that only 35 per cent of the population had access to relatively safe drinking water and only 32 per cent had proper sanitation. In other words, 1200 million people lacked safe drinking water, and 1400 million lacked sanitation.

The problem is by no means restricted to developing countries. In the OECD countries, for example, the percentage of the population served by a waste water treatment plant in 1983 ranged from 2 per cent for Greece and 30 per cent for Japan, to 100 per cent for Sweden.[3]

The United Nations International Drinking Water Supply and Sanitation Decade (1981–90) was launched to help rectify the scandalous situation in the developing countries. Its aim – that of providing everyone with drinking water and sanitation during the Decade – was nowhere near met. On the other hand, many millions of people were provided with proper water and sanitation and, in spite of population growth, a higher proportion of the population now enjoys safe water and sanitation than at the beginning of the Decade. According to the official estimate, access to a safe water supply in developing countries rose from 44 to 69 per cent over the Decade; proper sanitation was provided to 54 per cent of the population at the end of the decade, compared to 46 per cent at the beginning. In the cities, things are better than these averages imply; in rural areas, they are worse.[4]

Behind these facts lie some grim statistics. Water-borne disease, the complaint that dictates the "illfare" of most of the world's population, is ever-present in developing countries. According to the WHO, as many as 4 million children die every year as a result of diarrhoea caused by water-borne infection. Each could be saved by a simple packet of sugar and salts costing 7p. Better still, the risk could be removed completely if clean water were universally available, and sewage properly treated.

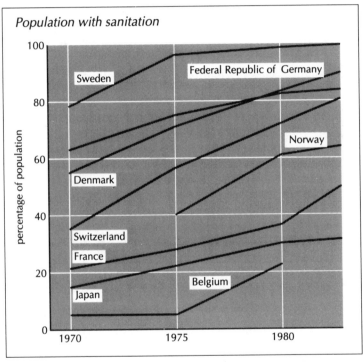

Proportions of European and Japanese populations with sanitation

Dirty water is responsible for more than diarrhoea. At the time of the United Nations Water Conference, held in La Plata in 1977, estimates were made of how many people were suffering from 30 of the world's major water-related diseases. Many of these diseases are not contracted from contaminated drinking water but are carried by vectors such as mosquitoes and snails that live and breed in irrigation water, river water and lakes. Water-borne disease, it was estimated, was affecting 400 million people with gastroenteritis, 200 million with bilharzia (caused by blood flukes carried by a water-borne snail), 200 million with filiariasis (threadworm), 160 million with malaria and 20–40 million with onchocerciasis (a disease caused by the nematode worm).

The overall effect on human health is virtually incalculable. Nevertheless, attempts have been made. According to one WHO official, dirty water was the cause of 27,000 deaths a day in 1986.[5]

Given these facts it is hardly surprising that the WHO should attach such great importance to clean water. "The number of water taps per thousand persons", declared Halfdan Mahler, WHO's Director-General in the early 1980s, "will become a better indicator of health than the number of hospital beds."

World Water Resources

There are about 1360 million cubic kilometres of water on the earth. If all the water on the planet – from the oceans, lakes and rivers, the atmosphere, undergound aquifers, and what is locked up in glaciers and snow – could be spread evenly over the surface, the earth would be flooded to an overall depth of some three kilometres.

More than 97 per cent of this water is in the oceans. The rest – about 37 million cubic kilometres – is fresh water but most of that is of little use since it is locked in icecaps and glaciers. Current estimates are that about 8 million cubic kilometres are stored in relatively inaccessible ground water, and about 0.126 million cubic kilometres are contained in lakes and streams.

How much is locked up in the water "larder", however, is much less interesting than how the larder is refilled. Only the renewable fraction of the world water cycle can actually be used for sustainable development. If ground water is used faster than it is replenished, then water reserves are being mined just as surely as is the coal that comes from deep

underground, and the oil and gas from under the sea bed.[6]
A great deal of ground water is currently being mined, notably
in Libya and the Midwest of the United States; Bangkok and
Beijing, to quote just two examples, are supplied largely from
mined groundwater reserves.

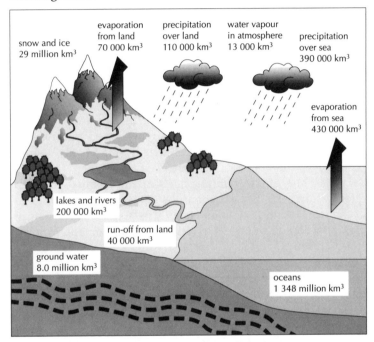

The global hydrological cycle

Rainfall and Evaporation

The rain that falls on the land averages some 725 mm a year.
In some places it rains gently throughout most of the year
and in others torrential rains occur for one or two months
a year, the rest of the year being almost rain-free. In some
places, such as the Atacama Desert, rainfall is effectively
zero; and there are places in tropical forests where there
are more than 5 metres of rain a year. In the world's arid
areas, where more than 600 million people live, rainfall is

less than 300 mm a year. Crops can be grown only under irrigation.

Rainfall is balanced every year by the world's run-off – the amount that flows out to sea via rivers and streams – and the amount that is evaporated from the land surface and released from the foliage of plants (together called evapo-transpiration). Far more rain is evaporated than ever reaches the sea via the rivers. Total rainfall on land areas amounts to some 110,000 cubic kilometres a year, of which about 70,000 cubic kilometres are evaporated. Thus only about a third of all rainfall ever reaches a river or stream – on average. In some places, where evaporation rates are high, the proportion of rainfall that is transformed into run-off is much less. In Africa as a whole, only about one-fifth of all rainfall is transformed into run-off. And in some of the world's drier river systems, only a small percentage of rainfall reaches a river.

This is one of the critical points in the chain for human intervention. This book is about the effects of water shortages in the world, and how to mitigate them. One of the secrets of making water go further is to reduce evaporation and increase run-off. And the basic way of doing this is to collect and use rainfall as near to where it falls as possible. The process is known as water harvesting, an activity likely to become just as important as food harvesting in the years to come. Indeed, in many places of the world, food harvesting is not possible – and never has been – without water harvesting first. As we shall see, many traditional societies survived by sophisticated water harvesting techniques that stored flash floods, the briefest of showers and even the night's dew for future use. These societies flourished in some of the most hostile environments known to the planet.

There is evidence that water harvesting used to be extensively practised in nearly all arid and semi-arid areas. But with the technology to drill deep boreholes, and pump up clean ground water, interest in water harvesting waned. The process has been almost forgotten in many parts of the world – but it will surely have to be revived. Water harvesting techniques are described in Chapter 10.

Run-off

The world's run-off is estimated at about 40,000 cubic kilometres a year from the land surface, excluding the polar zones. Evenly distributed, this would be enough to support a world population perhaps ten times larger than today's.

But neither run-off nor rainfall is by any means evenly distributed. Nearly a third of all rainfall falls on South America and the Caribbean, less than one per cent on Australia. And the Atacama Desert, probably the world's driest area, is in South America, one of the world's wettest areas. The coastal strip of the Atacama Desert rarely receives any rainfall at all. Its weather station, at Iquique, has recorded zero rainfall for continuous periods of more than five years.

Imbalances such as these cause many water shortages. So do more local imbalances within countries, and even within regions within countries. In Mexico, less than 10 per cent of the country's land area provides half the national run-off. The other 90 per cent of the country is dry indeed – though, overall, the country is relatively wet if judged solely on the basis of the nation's average annual rainfall.

Time, too, can deal an unfair hand. Areas that get adequate rainfall most years can get very little other years. Many parts of Europe suffered severe drought over the two-year period 1989–90, with 1989 rainfall down on average years between 11 per cent in parts of the United Kingdom to 60 per cent in Nice.[7]

The rainfall that can be relied upon every year, without fail, is therefore much lower than total rainfall. Planners use a figure of about 35 per cent of total rainfall to calculate the value of reliable rainfall. Even this does not tell the whole story, since where rainfall is zero for much of the year – as in many parts of the Australian outback, for example – even figures for reliable rainfall are meaningless. However much it rains, where there is no water at all for 10 or 11 months a year, survival becomes very difficult – if not impossible.

Calculating reliable supplies of run-off – as distinct from reliable rainfall – is even more complex, again because of the way rainfall varies over time. Much tropical rainfall, for example, comes in torrents during the monsoons. The spates produced by this form of run-off cannot be classed as reliable run-off because they are – literally – here today, gone tomorrow. Nearly two-thirds of reliable run-off is dissipated in this way – essentially in flash floods that provide little of value.

But this is the second point at which human intervention counts. It is possible, usually at great cost, to increase the volume of reliable run-off by building reservoirs for water storage and using canals to move water from water-rich areas to water-poor ones. These solutions determine, as the United Nations Educational, Scientific and Cultural Organization (Unesco) has put it, whether water on the ground becomes "a productive resource or a destructive hazard".[8] Large-scale solutions are described in Chapter 8; because of their expense, and their environmental and social costs, they are becoming increasingly unpopular. There are small-scale alternatives, however. These are described in Chapters 9 and 10.

The Beginnings of Water Stress

So where do water problems really start? Experience in Europe suggests a few rules of thumb. Countries that use less than 5 per cent of total run-off have few water management problems. Countries that use 10–20 per cent usually have fairly major water problems. Above 20 per cent, water supply becomes a major national issue and may even prove to be the limiting factor in economic development.

In developing countries, of course, these figures would be unduly pessimistic were it not for one important factor: the drier a country, the greater its need for irrigation water. And irrigation is by far the most intensive use of water known. Some of the world's driest countries use nearly all their total run-off just for irrigation. Without it, they would not be able to grow even the crops they do.

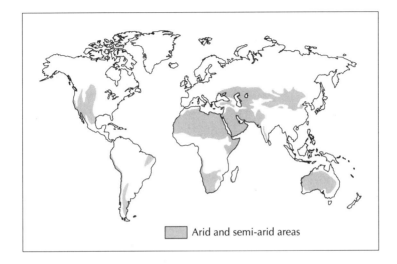

Arid and semi-arid areas

Three areas of the world are particularly short of water: Africa, the Middle East and South Asia. These form the bulk of the arid and semi-arid areas of the world. Other dry areas include the south of North America, limited areas in South America and parts of Australia. In each of these areas, the run-off available per head of population is either low or very low. For example, India is already using half its total available run-off, and drawing half as much again from groundwater resources. By the year 2025, it is estimated, demand will reach 92 per cent of India's total annual freshwater resources.

Indian statistics provide a vivid illustration of how the water is used. Industry accounts for only about 1.4 cubic kilometres of water a year. Domestic demand is double that. Livestock and power stations each use three times as much. But irrigation uses 360 times more, accounting for more than 97 per cent of all water use.

Water-use rates may already be greater than is sustainable. In many areas of the country the water table is falling, presaging the onset of groundwater drought. Such profligate use of ground water is not, of course, confined to India. In the Beijing area of China, so much ground water was pumped up recently that the water table fell by four metres in just one year.

In Maharashtra State, in India, the sugar factories have been responsible for deepening many boreholes to improve their water supplies. The effects have generally been short-lived. In the village of Tasgaon Taluk, a new water scheme was commissioned in November 1981, designed to supply 50,000 litres a day to the sugarcane factories. It ran dry a year later, and three boreholes were then sunk, near the well, to a depth of 60 metres. These also ran dry a year later. Water now has to be brought in by tanker from a supply more than 15 kilometres away.

Another result has been that more than 2000 wells have run dry in the Tasgaon Taluk area. In the state as a whole, where once there were only 1810 villages with water problems, 23,000 villages now have no source of drinking water.

To many of the local inhabitants, this looks like the effects of drought. It is not. It is a far more serious affair, one that is permanent rather than temporary, one that threatens to get worse rather than better.

History: Some Cautionary Tales

It is hard to underestimate the importance of water to society. The earliest forms of civilization developed along the banks of such rivers as the Tigris and the Euphrates, the Nile and the Indus. Irrigation was the economic force that drove these societies, the rivers that supplied their irrigation water their one essential requirement for survival.

Water management in these societies was no mean achievement. Through complicated series of dikes, canals, weirs, reservoirs and lifting devices these early civilizations managed to transform the "hydrological chaos" of the great river valleys into the well-organized and regularly watered fields, meadows and gardens which, in the Tigris-Euphrates valley, were known mythologically as the Garden of Eden. By 2500 BC the Harappans of Mohenjodaro not only had brick-lined wells but had drainage channels constructed from burnt brick to take away foul water. But their success in controlling domestic water supplies was not matched by their long-term ability to manage irrigation.

The organization of irrigation is a complex business, and one which some historians claim exerts a profound influence on social evolution. Certainly it is true that rulers in past civilizations were able to use control of irrigation works as a means to political power. Whether, as some claim, societies that rely extensively on irrigation are always characterized by rigidly centralized control is another matter.

Mesopotamian civilization collapsed, in the end, because it was unable to manage its critically important water resource adequately. Its soils became salinized and waterlogged, and its water supply systems silted up. Today, none of this is surprising because we understand better the enormous problems with which the Mesopotamians were faced: their annual flood came in April–June, too early for the summer crop, their rivers carried huge amounts of sediment, and the fine texture of their soil and the slope of their valley – only 1 in some 26,000 – led inevitably to waterlogging and salinization.

The process of decline was a long one. As a result of poor drainage, groundwater levels inevitably rose over the centuries, carrying with them the salts that were eventually to poison Mesopotamian agriculture. Salinity, initially, did little more than force people to move to new land, although farmers experimented with fallow land techniques, crop rotation and the use of salt-resistant crops. Wheat and barley were the principal crops, but of these wheat was much the less salt-tolerant. The proportion of wheat grown in southern Mesopotamia fell from some 50 per cent in 3500 BC, to 15 per cent in 2500 BC and eventually to 2 per cent in 2000 BC Barley yields declined in almost parallel fashion: from 2600 litres per hectare in 2400 BC to only 1000 litres per hectare in 1000 BC.

Today, about 80 per cent of the area is affected by salinization, and farmers have abandoned at least one-third of it. Strangely, the factors that caused the decline of the world's first great civilization were not fully understood until well into the 20th century. Even after World War II, irrigation schemes were still being installed without proper drainage.

Water was equally important to many other early civilizations. In Egypt irrigation was being practised by 3400 BC, and in the New World the Incas in Peru were using it in 1000 BC. The Greeks were the first to introduce long-distance, high-pressure water supply pipelines. Water management was also taken seriously by the Romans, who exported their ideas to many parts of the Middle East. In Rome itself water management became a fine art, with 14 aqueducts, with a total length of more than 500 kilometres, bringing water to the city and to its ornate and elaborate systems of public baths. Water distribution within the city was through lead pipes, a system that was still being used in Europe after World War II. But water problems were again implicated in the decline of the civilization, at least in the Roman Empire in North Africa, where massive soil erosion, caused by inappropriate agricultural techniques, led to the collapse of the Roman "bread basket".

Greek civilization had earlier suffered a similar fate, as Pliny himself observed:

> In comparison to those [old] times, like a body of whom because of a wasting disease only the bones are left, the fertile and soft soil is everywhere eroded and only the sterile skeleton is left. But in those [old] times, when the land was still undamaged, its mountains were high and covered with earth and likewise its plains, which are now called stony fields, consisted of fertile soil. On the mountains there grew plenty of trees of which today there are still distinct traces.
>
> Plato, Critias

In the early days of civilization, water was usually the principal challenge. Cities such as Jericho, Ur, Memphis, Babylon, Athens, Carthage, Alexandria and Rome relied completely on their water engineers, without whom they could never have existed. They were the people who found the relevant water resources, brought them to their cities, arranged for storage and distribution, and made at least some attempt at proper sewage disposal.

The contrast with today could hardly be greater. Water shortages in the developing countries attest to the casual way in which water has been taken for granted in the development plans of most countries. And the pollution problems of the developed countries reflect Western attitudes to their water resources – resources they take for granted and use mainly as a means of flushing away unwanted wastes. Today's water engineer attracts none of the respect of the water managers of the past – the people who, almost literally, were responsible for lubricating the wheels of ancient civilizations.

The art of water management was essentially lost during the Middle Ages. Late in the nineteenth century scientific discoveries once again made clear how important water was to human life, and many water-borne diseases were banished from industrial society. Though we have not lost sight of this, we have not yet found the means to provide the world's citizens with the clean water and efficient waste disposal systems that our ancient ancestors would have insisted upon.

Notes

1 Disease from polluted water is by no means the prerogative of developing countries. Poor management, lack of investment and slack controls are leading to increasing numbers of cases of "water poisoning" in industrial countries, where low water quality standards, both inland and along the coastline, have become a scandal in many places in Europe, North America and Japan. In Canada, for example, it is estimated that the health of as many as a million people may be being affected by the consumption of contaminated well water.

2 In England. In ancient Rome, the first sewage system was developed jointly with water supplies, and the first closed sewage system, the *Cloaca Maxima*, was built in 500 BC. Eventually all sewage was disposed of through closed systems, regularly flushed by overflows from the water supply system. Frontinus wrote: "Even the water

spilled from the supply system is used. The causes of the unhealthy climate are being flushed, the sight of the streets is tidy, cleaner is the air we breathe, eliminated is the atmosphere which earned the city at the time of our ancestors a poor reputation." In most parts of the world, sewage disposal systems still lag behind those of the Romans.

3 Michael Meybeck, Deborah Chapman, and Richard Helmer, *Global Freshwater Quality* (Oxford, Blackwell, for UNEP and WHO, 1989).

4 Some confusion surrounds these figures. For example, the definition of an adequate supply of safe water at the beginning of the Decade was taken to mean that a source was available within 200 metres. By the end of the Decade, a number of countries had changed their accounting system; India, for example, classified any good quality water supply within a one and a half-hour walk as "adequate". The apparent success of the Decade may therefore be due at least partially to creative accounting methods. See *Levels of safe water and sanitation services 1980 and 1990, all developing countries.* (UNDP Information Paper, 1990.)

5 S. Barabas, "Monitoring Natural Waters for Drinking Water Quality", *World Health Statistics Quarterly*, 39(1), 32–45.

6 Renewability can be a relative term. Coal and oil are created, though only slowly compared to current rates of use, and are thus at least theoretically renewable. Furthermore, there are possible replacements for fossil fuel energy such as hydropower, wind power, tidal power and nuclear power. There is no such possible replacement for water, and therefore the limitations to development that arise as a result of water shortage are more fundamental than those implied by shortage of the currently fashionable energy forms.

7 Mick Hamer. "The Year the Taps Ran Dry", *New Scientist*, 18 August 1990.

8 W. E. Cox (ed.), *The Role of Water in Socio-Economic Development* (Paris, Unesco, 1987, *Studies and Reports in Hydrology 46*).

Chapter 2

Water Needs

The energy crisis of the 1970s will take a back seat to the water crisis of the 1980s and 1990s.

United States Department of Agriculture
The Yearbook of Agriculture, 1981

Like coal, oil, soil or iron ore, water is a natural resource. But there are many ways in which water differs from other natural resources. First, it moves. Second, its total quantity on the earth is fixed, and can be neither increased nor decreased. Thirdly, water is essential for human survival.

Man's biological need for water is modest. A dozen or so cupfuls a day are all that are required for survival. Even so, there are many areas of the earth where even this requirement is difficult to meet. Rainfall in many desert regions is limited to a few millimetres a year, and this often falls at unpredictable times during the space of a few isolated days. Survival in these regions is impossible unless water is imported.

Household Needs

Biological survival, however, is not the issue in today's water-stressed world. Water is required for household needs, for industry and for agriculture. Household needs – drinking, washing and cooking – could be adequately met everywhere in the world by less than 100 litres per person per day, roughly the amount used for an average shower. Of this, only one litre a day is required for drinking. A hundred litres a day is the equivalent of about 35 cubic metres per person a year. Even the most arid countries would, in theory, have little difficulty in providing this volume from run-off. Malta, which is one of the driest countries in the world, has a run-off equivalent to 70 cubic metres per person per year. To water-rich countries, domestic use is tiny. Canada, for example, could supply 121,930 cubic metres per head per year.[1]

There is thus no global Malthusian constraint on water availability for domestic use – though there are many local constraints. If every man, woman and child on the planet were provided with 100 litres of domestic water a day, the water bill for a population of 5 billion people would be 180,000 billion litres a year, or 180 cubic kilometres.

This compares with a global run-off figure of 40,000 cubic kilometres a year. As mentioned in Chapter 1, global run-off is not the same as useful, reliable run-off. Most of the rainfall on the land areas of the earth – which averages less than a metre a year – is almost immediately lost in floods, and can be considered neither stable nor reliable. This portion of run-off amounts to about 26,000 cubic kilometres a year. Of the remaining 14,000, some 5000 falls on virtually uninhabited areas unsuitable for human settlement. This leaves about 9000 cubic kilometres a year, the equivalent of about 1800 cubic metres per person per year.[2]

Even so, the theoretical domestic water requirements for the present population amount to only about 2 per cent of what is reliably available. In theory, the Amazon alone – with an annual flow of nearly 6000 cubic kilometres – could supply the domestic water needs of a world population more than 30 times as large as it is now.

Reliable figures for what is actually used are hard to come by. While statistics are reasonably accurate for many developed countries, only the vaguest estimates can be made of what is used, for example, by a nomadic pastoralist in the Sahel, a charcoal burner in Amazonia, or a sheep drover in the Australian outback.

According to the World Resources Institute, domestic use in 1986 varied from some 200 cubic metres per person per year in the United States to only some 7 cubic metres in Oman.[3] But of course much less water is used in many arid and semi-arid areas, where one tap or well often supplies hundreds of people. In 1983, for example, less than 40 per cent of the world population had access to a safe source of drinking water. This figure was probably not much higher even at the beginning of the 1990s.

The data that do exist for developing countries reveal more about the hardship involved in obtaining drinking water than they do about the statistics of water use. A survey carried out in the early 1970s in East Africa, for example, showed that the average time spent collecting drinking water was 46 minutes a day but could rise to 4 hours. The women carried about 15 litres on each journey.[4]

Such effort does not encourage conspicuous consumption. A Nigerian survey has shown that per capita use in an area where water had to be carried was 21 litres a day, while nearby urban dwellers with taps used 82.1 litres.[5] The Intermediate Technology Development Group found that in Swaziland per capita use was normally about 5 litres a day, rising to 13 litres for families which could afford to pay for delivered water. Tap users averaged 30–100 litres a day.[6]

Not all this water is used for drinking. But drinking and cooking water make a substantial hole in a daily allowance of five litres, leaving precious little for washing either people or dishes.

In 1977 the global use of domestic water was estimated at 100 cubic kilometres a year – the equivalent of about 20 cubic metres per person per year.[7] As a result of population growth, and of ambitious plans to increase the availability of safe drinking water and sanitation throughout the developing countries, this was expected to rise more than nine fold by the end of the century, to 920 cubic kilometres a year, more than 150 cubic metres per head of the projected population. Much of this increase would be for sewage disposal, which is a water-intensive process, and is really a municipal rather than a strictly domestic use. The figure of 150 cubic metres per head per year is equivalent to 400 litres per head per day, four times as much as the basic minimum proposed at the beginning of this chapter. The lack of success of the International Drinking Water Supply and Sanitation Decade, which ended in 1990, now makes these figures look wildly optimistic.

Although the issue of supplying drinking water and sanitation is not the central theme of this book, it is critically important. A supply of unpolluted drinking water, and the sanitary disposal of human wastes, are fundamental to health. Furthermore, lack of convenient water supplies puts great stress on families in developing countries, where it is the women and children who are mainly responsible for water carrying. Not only is this arduous, back-breaking, and repetitive work but it also occupies considerable portions of the days of many children. Even in towns, women and children spend long hours queuing for water at a tap that has to supply hundreds of families. Schooling and education suffer when children are forced to fetch and carry water.[8]

That said, family demands on water supplies are not large when viewed in the context of water as a scarce resource. Domestic use is small compared to industrial use, even in some developing countries; and it is almost negligible in comparison to the volumes used for irrigation.

The Needs of Industry

Household and even municipal water needs are only a small part of the water supply problem. Globally, industrial water use is at least twice domestic use. In addition, vast quantities of water are used by power stations as cooling water. Although this water is not "used" – it is returned directly to the river from which it is taken, unaltered except for a temperature increase of a degree or two – it must still be subtracted from available run-off. Power stations must have a reliable supply of cooling water (a condition that is increasingly hard to meet in some areas). This means that the water power stations use cannot be diverted upstream for other purposes. Cooling water is therefore a drain on water resources.

Taken together, these two uses – industrial use and cooling – amount to more than four times domestic water use, and they must be supplied from reliable sources of run-off. The same is true of water used for livestock, which is about half that of domestic use. In broad terms:

GLOBAL WATER USE BY SECTOR PER YEAR

domestic	100 cubic kilometres
industrial	200 cubic kilometres
cooling	225 cubic kilometres
livestock	40 cubic kilometres
TOTAL	565 cubic kilometres

These figures, which are from a 1977 estimate,[9] imply per capita values for domestic, industrial, cooling and livestock use of about 20, 40, 45 and 8 cubic metres per head per year – in round terms, an annual total of some 120 cubic metres per person. Compared to what is available, even these figures are relatively small, amounting to less than 10 per cent of reliable run-off. So far, it is tempting to conclude, there is no real problem.

The Issue of Quality

Such a conclusion would be misleading. One reason is that global figures hide local and regional problems. In Finland, industry accounts for about 85 per cent of all the water used. The United States, Canada and Poland all use about 40 per cent of their total use as cooling water. Belgium uses as much as 47 per cent for cooling and a further 37 per cent for industry. Even Nicaragua and Barbados use 45 and 35 per cent respectively for industry, though in some developing countries industrial water use is lower than domestic water use.

Inevitably, such large-scale industrial use of water entails severe problems of water quality. So does the domestic use of water: in many developing countries river pollution from raw sewage reaches levels that are many thousand times higher than recommended safe limits for both drinking and bathing.

The problem of water quality actually becomes the key issue in water use. The reason is that what industry uses and what industry needs to decontaminate the water it has polluted are different matters. Thus it is estimated that every year some 450 cubic kilometres of waste water are allowed to flow into rivers and streams. But a further 6000 cubic kilometres of water are then needed to dilute and transport this contaminated water before it can be used again – roughly two-thirds of total reliable run-off. Seen in this light, water availability looks much more problematic. Indeed, it is estimated that the entire stable global river flow will be required for pollutant transport and dilution by the year 2000.[10]

There can be no clearer indication of the need for the recycling of water by industry. The situation is already critical, with many rivers flowing into the sea uncleaned, and depositing into the oceans a large portion of the world's industrial – and domestic – debris. Whereas in pre-industrial society it was considered as a rule-of-thumb that a polluted stream or river cleansed itself every few kilometres, today's rivers and streams are often polluted from source to estuary. In this sense, the water crisis is already upon us.

Water quality has become the major environmental issue in many industrialized countries. Even in water-rich Canada, which has nine per cent of the world's fresh water, there are local water shortages and widespread contamination of both surface and ground water. Canada has already introduced pricing and metering schemes to reduce water use. A recent report by the Science Council of Canada has called for a new approach to water management based on sustainable use.[11] In the United States, as in most industrial countries, there has been widespread industrial contamination. Peter Rogers writes:

> In the San Gabriel Valley of California, for example, thirty-nine wells that supplied water to thirteen cities had to be closed in

1980 when they were found to be polluted by high concentrations of trichloroethylene (TCE), an industrial solvent and degreaser. Seven municipal wells and thirty-five private wells near Atlantic City, New Jersey, had been closed by March of 1982, after wastes from a chemical dumpsite were found to have seeped through the sandy soil into ground water. In Bedford, Massachusetts, four wells providing 80 per cent of the town's drinking water were closed in 1978 when they proved to have been contaminated by high concentrations of dioxane and TCE, among other toxic chemicals. By 1979, drinking water in a third of the communities in Massachusetts had been found to be contaminated.[12]

Public alarm over water quality has produced change in developed countries, and there have been improvements in quality in many major rivers – though not in ground water – over the past 20 years. All the indications are, however, that the change has come too late and has not gone far enough. In developing countries, progress has been virtually non-existent.

Industrial pollution is less common – though still severe near most of the major urban areas – in many developing countries. Even so, it is estimated that 70 per cent of India's rivers are polluted with industrial waste. Friends of the Earth in Malaysia reports that the situation there is, if anything, even worse. Even in 1979, many of the major rivers in peninsular Malaysia were officially declared dead as a result of pollution mainly from oil-palm and rubber effluents, and other industrial wastes. These rivers no longer sustained fish, shellfish, crabs or shrimps. They were unfit for both drinking and bathing.

Domestic pollution from untreated sewage poses acute health problems along nearly every major river in the developing world. To quote the World Resources Institute:

Out of India's 3119 towns only 217 have partial (209) or full (8) sewage treatment facilities. The result is severely contaminated waters. A 48 kilometre stretch of the Yamuna River, which flows through New Delhi, contains 7500 coliform organisms per 100 millilitres of water before entering the capital, but after receiving an estimated 200 million litres of untreated sewage every day, it leaves New Delhi carrying an incredible 24 million coliform organisms per 100 millilitres. Industry is no better. The same stretch of the Yamuna River picks up 20 million litres of industrial effluents, including about 500,000 litres of "DDT wastes" every day.[13]

The safe level of coliform concentration for drinking water is about 100 organisms per 100 millilitres of water. Since the average human excretes about 2 billion such organisms a day, it is easy to see how quickly water containing untreated sewage can become contaminated. Coliform levels above 3 million per 100 millilitres have been reported in both the canals of central Jakarta and in Nigerian ponds. A count of 7.3 million has been recorded in the river downstream of Colombia's capital.

While the problem of sewage contamination of rivers is confined mainly to developing countries, raw sewage is commonly pumped into shallow waters off the coasts of most countries in the world. Many of Europe's beaches are seriously contaminated and should be declared unfit for swimming. Even in Canada, beaches near Toronto had to be closed three summers running during 1984–86. The European Community has issued directives on acceptable water quality on beaches which some countries, notably the United Kingdom, have ignored. One of the UK's arguments in this fierce political battle was that the definition of a beach should be applied only to places where 500 bathers were actually in the water at any one time or where there were more than 1500 bathers per linear mile of beach. This definition excluded not only all Welsh beaches from the category of "beach" but also excluded Blackpool – the United Kingdom's best-known seaside resort.[14]

Water quality in the rivers of the developed countries suffers from many other contaminants. Chemical pollution is common, and levels of nitrates from agricultural fertilizers are high. While water treatment can cleanse such water to some extent, nitrates are difficult to eliminate and treatment is expensive – far too expensive for most developing countries.

The issue of water quality thus has considerable bearing on water supply. A reliable run-off of 1800 cubic metres per head of population per year may seem ample; but when much of that water is fit neither for drinking nor bathing, the whole concept of reliable run-off is called into question. "Reliable" is not a word that accurately describes the muck that flows down most rivers.

The Needs of Agriculture

Irrigation is by far the biggest user of water, and also the most rapidly expanding.

Plants use large quantities of water during their growth. Under dry conditions, it takes about 1000 cubic metres of water to produce one tonne of plant growth. Where rainfall alone is insufficient to meet plant needs, irrigation is required and the volume of water needed rises dramatically – partly because of the nature of the irrigation process and partly because of human inefficiencies.

Peter Stern puts it graphically:

> . . . if all the water consumed in a month by a rural community of 1000 people with 250 cattle and 500 sheep and goats were used for irrigation, this would provide two irrigations a month to an area of about one quarter of a hectare.[15]

The amount of water used for irrigation has increased 10 times this century, and elaborate plans are still being made to extend irrigation to more and more areas.

The watering of crops currently uses something like 3300 cubic kilometres of water a year – roughly six times the requirement for industrial and domestic uses. This amount is not supplied entirely from stable run-off. An increasingly large proportion comes from pumped ground water, and in many places irrigation must wait for the seasonal supply of water, frequently that brought by the monsoon. In desert areas, irrigation depends principally on the flash floods used in wadi irrigation which may occur only two or three times a year.[16]

Irrigation is a wasteful process. Less than half of all irrigation water ever reaches the crop it is designed to water. Sometimes the water returns to an underground aquifer or nearby river, and little is wasted. More often, though, it is polluted by salts or nitrates first. When it finds its way back to the water system, it spreads its pollution further, requiring large quantities of unpolluted water to transport it away and dilute it. Thus waste is magnified, usually in places where water waste can least be tolerated. It is one of the more perverse laws of nature, of course, that only those countries that are by definition short of water need to irrigate. Water-rich areas can rely on rain-fed agriculture.

Unless land is irrigated carefully, irrigation can produce a great deal of environmental damage. Yet it can also bring enormous benefits, when properly managed. Irrigated crops have higher potential yields than many rain-fed ones. Irrigation extends the area of land that can be cultivated since most irrigation is practised in areas that are too dry for rain-fed farming. And irrigation often allows multiple crops to be taken in the same year, thus further increasing productivity.

Finally, irrigation can provide security to the farmer who was previously dependent on unreliable seasonal rains, encouraging him to use higher-yielding varieties and more costly inputs such as fertilizers and pesticides on which he might not dare risk spending money if his water supply were unreliable. For all these reasons, irrigation is a force to be reckoned with. Although irrigation uses the lion's share of water resources, it is critically important in helping developing countries grow the food they need.

Irrigation accounts for most of the water used in many developing countries. In Egypt, for example, more than 98 per cent of all the water used is applied to crops. Domestic water use is less than 10 cubic metres per person per year but agriculture accounts for the equivalent of more than 950 cubic metres per person per year. Thus Egypt's total per capita use of water is only marginally less than that of water-rich Canada; India's and China's is half as much, with 90 percent of their water use going to crops. The United States, which uses 40 per cent of its water for agriculture, uses a princely 2000 cubic metres of water per person per year.

Not surprisingly, there are signs that the explosive growth period of irrigation is drawing to a close. Water shortages are one of the main reasons. As Sandra Postel puts it:

> In several parts of the world, water demands are fast approaching the limits of resources. Many areas could enter a period of chronic water shortages during the nineties, including northern China, virtually all of northern Africa, pockets of India, Mexico, much of the Middle East, and parts of the western United States. Where scarcities loom, cities and farms are beginning to compete for available water; when supplies tighten, farmers typically lose out.[17]

While much of the world's irrigation water is fed from rivers, lakes and reservoirs, an increasing proportion is now pumped up from ground water. Underground aquifers are fed from water trickling down through the soil and, occasionally, by the disappearance of streams and rivers underground. Most are in continual circulation – although some, the so-called fossil aquifers, appear no longer to have renewable supplies.

There is no reason why ground water should not be used for irrigation – or, indeed, for industrial or domestic supply.[18] But there are many complications. The first is that ground water is renewed much more slowly than other water sources. For example, on average, the atmosphere's moisture is renewed every eight days, stream water is renewed every 16 days, soil moisture is renewed annually, swamp water every 5 years and lake water every 17 years. Ground water is renewed only once every 1400 years.[19]

Providing ground water is not extracted faster than it is renewed, its use can be considered sustainable. All too often, though, extraction rates now exceed renewal rates. This is unsustainable development, and nature complains loudly about it. When water is overpumped, water tables fall, land subsides, and saline water can infiltrate coastal aquifers. Pumping the water becomes increasingly expensive as levels drop. In California, China and India, many irrigation projects have been halted for these reasons. Irrigation water pumped from underground aquifers must therefore be treated as part of run-off in any calculation of overall water use.

Adding up the Figures

The basic addition is simple: 565 cubic kilometres for domestic, industrial, cooling and livestock use, plus 3300 cubic kilometres for irrigation. As near as makes little difference, the answer is 4000 cubic kilometres a year, equivalent to 44 per cent of total reliable run-off.

In 1940, total water use was about 1000 cubic kilometres a

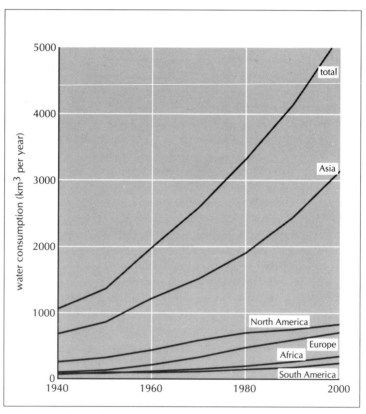

Global water use by continent, 1940–2000

year. It had doubled by 1960, and doubled again by 1990. The crux of the water crisis is that it probably cannot double again because geographical constraints would make it nearly impossible to use the equivalent of 88 per cent of total reliable run-off on a global basis. Countries that do use water at these rates experience chronic problems of water shortage during dry periods, and of rapidly declining water quality.

WATER USE BY CONTINENT (km^3 per year)

	1950	*1960*	*1970*	*1980*	*1990*	*2000*
Africa	56	86	116	168	232	317
N. America	286	411	556	663	724	796
S. America	59	63	85	111	150	216
Europe	93.8	185	294	435	554	673
Oceania	10	17	23	29	38	47
TOTAL	1360	1982	2594	3316	4138	5189

Source: Michael Meybeck, Deborah Chapman, and Richard Helmer, *Global Freshwater Quality* (Oxford, Blackwell, for UNEP and WHO, 1989).

The Swedish hydrologist Malin Falkenmark estimates that the needs of the temperate zone industrialized countries can be met by between 150 and 900 cubic metres per person a year. The irrigated semi-arid countries, by contrast, need about five times as much – between 700 and 3500 cubic metres per person a year. And irrigated semi-arid industrialized regions, such as the lower Colorado basin in the United States, need even more: 2700 to 7000 cubic metres per person a year. "Need" is perhaps not the right term. With care and recycling, some semi-arid, semi-industrialized countries have reduced their water use to relatively low levels. Thus Israel uses 447 cubic metres of water per head per year, and South Africa only 404.[20] However, Israel does not strive for food self-sufficiency but instead exports many high-priced, luxury food items.

A few sample figures illustrate the predicament. Libya has 190, Kenya 720, Egypt 1200, India 2430 and China 2520 cubic metres per person a year. All these are irrigated semi-arid countries, with a requirement of 700–3500 cubic metres per person a year. Even in 1985 some of them could not meet their needs adequately. Is it surprising that development plans – which frequently pay scant attention to water resource issues – often go so badly wrong in so many of these, and other similar countries? Is it surprising that land degradation, the pollution and exhaustion of underground aquifers, and desertification are rampant in these areas?

The rising need for water has two components: one is that more and more people use water, and the other is that they use more of it than they used to. In 1940 per capita water use was below 400 cubic metres a year; it is now at least double that at 800 cubic metres a year. Over that period the world population has also doubled. It is this combination of increased use and increased population that produces water stress.

The main parameters of this equation, and its implications for the future, are examined in more detail in Chapter 5. The next two chapters deal with other important factors that promise to exacerbate the water crisis considerably: the influence of a changing climate on water supplies, and the dessication of the earth's surface that is resulting from human activity.

Notes

1 1985 figures, taken from Adrian T. McDonald and David Kay, *Water Resources: issues and strategies* (London: Longman Scientific and Technical, 1988).

2 Two sources quote a figure of 9000 cubic kilometres a year: Robert P. Ambroggi, "Water", *Scientific American*, September 1980; and World Resources Institute, *World Resources 1986* (New York: Basic Books, 1986). More optimistic estimates of stable reliable run-off put the per capita figure considerably higher. Thus if reliable run-off were 14,000 cubic kilometres a year, per capita availability would be more than 2800 cubic metres a year for a population of 5 billion.

3 World Resources Institute. Op. cit.

4 G.F. White et al., *Drawers of Water: Domestic Water Use in East Africa* (Chicago: University of Chicago Press, 1972).

5 Akintola et al., "The Elements of Quality and Social Costs in Rural Water Supply and Utilization", *Water Supply and Management*, 4, 275–82, 1980.

6 ITDG, "Water for the Thousand Million", in United Nations, *Proceedings of the United Nations Water Conference* (Oxford: Pergamon Press, 1978).

7 M.I. L'Vovich, "World Water Resources Present and Future", *Ambio*, 6(1), 13–21.

8 Irene Dankelman and Joan Davidson, "The Invisible Water Managers", in *Women and Environment in the Third World* (London: Earthscan, 1988).

9 M.I. L'Vovich. Op. cit.

10 M.I. L'Vovich. Op. cit.

11 Science Council of Canada, *Water 2020: sustainable use for water in the 21st century*. Report 40, 1988.

12 Peter Rogers, "The Future of Water", *The Atlantic Monthly*, July 1983.

13 World Resources Institute. Op. cit.

14 Adrian T. McDonald and David Kay, *Water Resources: issues and strategies*. Op. cit.

15 Peter Stern, *Small-scale Irrigation* (London: Intermediate Technology Development Group, 1979.)

16 FAO, *Spate Irrigation* (Rome: FAO, 1989).

17 Sandra Postel, *Water for Agriculture: facing the limits* (Washington, DC: Worldwide Institute, 1989, Worldwatch Paper 93).

18 Much of the world's drinking water is, in fact, pumped up from underground. In the Soviet Union, for example, 60 per cent of towns are supplied exclusively from groundwater, 25 per cent have mixed supplies and only 15 per cent rely on surface water. There is good reason for this practice: groundwater is often, though not invariably, less contaminated than surface water; and it can often be found closer to hand than a reliable surface water supply.

19 V.I. Korzoun and A.A. Sokolov, "World Water Balance and Water Resources of the Earth", in United Nations, *Water Development and Management Proceedings of the United Nations Water Conference* (London: Pergamon Press, 1978).

20 World Resources Institute and the IIED in collaboration with UNEP, *World Resources 1988–89* (New York: Basic Books, 1988).

Chapter 3

Water and Climate

Let not a single drop of water that falls on the land go into the sea without serving the people.

Parakrama Bahul, King of Sri Lanka
(AD 1153–86)

The climate is an easy scapegoat for the ills of the earth. When famine strikes in India, the African Sahel or Peru, it is less risky politically to blame drought or climatic change than the incompetence of politicians or aid experts. And, in one sense, it is also more correct: were these areas blessed with more equitable climates, famine would be less likely.

Yet the evidence that the arid and semi-arid areas of the world are undergoing climatic changes of the scale that would be required to produce major famines is slim. Droughts are another matter. Droughts have always occurred in low-rainfall areas from time to time. As one group of experts put it, "The semi-arid areas are normally drought-prone".[1]

Societies that live in these areas have evolved mechanisms for coping with drought. The pastoralist who herds more cattle than he apparently needs is not playing the role of a greedy livestock supplier: he is planning for the future, taking out an insurance against the certainty that there will come a time when lack of rain prevents his grass from growing and his animals from thriving. When that happens he cashes in his insurance policy, selling his cattle for a good price in a market where demand outstrips supply. He invests again when the rain and the grass return.

Similarly, arable farmers in arid areas have a range of

techniques for increasing their security and dealing with drought. They grow mixed crops in the same field, alternating drought-resistant ones with others which may be more profitable but which are also more likely to succumb to dry periods; they mulch their fields to conserve what little moisture they do have; they dig earth barriers or build stone ones along the contours of the land to funnel water to cultivated areas; and they mix crop production with the raising of both animals and trees as an insurance against drought. These and other traditional techniques may hold the key to success in overcoming impending water shortages in many parts of the tropical world. They are dealt with in more detail in Chapter 10.

But if the climate is not – at least as yet – undergoing radical change, and droughts are not becoming markedly more severe or more frequent, why do such strategies apparently now fail? Why do peoples who used to know how to survive the hard times now succumb to them? There are many answers: there are, for example, now many more people – it is estimated that the rainfed arid and semi-arid areas of the world now support more than 1 billion people, one-fifth of the world population; many of them have been made much more vulnerable to climatic variability through a series of political decisions that have favoured cash crops over staple food production, urban dwellers over rural societies, and imported food over domestic produce.

These decisions have impoverished small-scale farmers in developing countries all over the world, forcing them to work more marginal land and to adopt land-use techniques that strip the land of its natural fertility. Peasant farmers have always had to walk a tightrope over disaster; now they must do so with their arms tied, and nothing to balance with. There is no safety net, and tragedy strikes more often and more seriously.

In this situation, marginal changes in climate, more frequent droughts, and reduced levels of rainfall will expose the basic instabilities already in existence: their impacts will be multiplied many times by the new vulnerability of rural society.

The Evidence for Drought

The 1970s and 1980s were unquestionably decades of drought. Lower than average rainfall, reduced productivity, and – above all – rural hardship and suffering were reported from California, Mexico, Peru, India, China and, of course, throughout much of Africa.

Famine itself, however, was restricted mainly to Africa. It is probably not coincidence that Africa has the most difficult climate of all the continents. Low rainfall and high insolation combine to provide Africa with the lowest ratio of run-off to precipitation of any continent; in other words, more rain evaporates in Africa, before it ever reaches a stream or river, than elsewhere. This is why farmers and cattle herders have such a hard time in the semi-arid regions (it is also why there are no canals in Africa). Total river run-off in Africa is only about 20 per cent of rainfall; in Europe it is more than 40 per cent.

After the droughts of the 1970s and 1980s, the African climate was subjected to rigorous analysis. This suggested that rainfall declined over the two decades in the Sahel and surrounding countries. But the decline, statistically at least, was marginal, and the data on which this conclusion was based were insufficient to establish any long-term trend with certainty. The Sahel has been dry for millennia, and some years have always been drier than others. Records go back only a few decades, and evidence from other periods must be adduced from indirect sources, such as stories of past droughts, geological evidence and archaeological remains. Such studies suggest that major droughts have tended to occur in the Lake Chad area, for example, approximately every 80 years since the seventeenth century.

In 1980, the World Meteorological Organization believed that nothing abnormal was occurring in Africa: ". . . the droughts of the 1970s," it claimed, "are normal to the climate, in the sense that they have occurred before, and presumably will occur again."[2] However, rainfall was below average in the Sahel over much of the period starting in the 1960s and continuing, with only a few short-lived exceptions, until today. According to Michael Dennett, of the University of Reading in the United Kingdom, "The Sahel's rainfall from 1974 to 1983 was about 5 per cent less than in the 1931 to 1960 period".[3]

This has apparently been enough greatly to reduce the amount of water flowing in the Senegal, Niger and Chari rivers in West Africa. Lake Chad itself has been reduced to one-third the area it occupied in 1963, with its northern reach now drying up annually. Even so, levels appeared no lower than they had been at times in the fifteenth, sixteenth, eighteenth and nineteenth centuries. In the 1820s and 1830s the Nile is reported to have flowed only weakly and Lake Chad itself to have virtually dried up.[4]

None of this may be abnormal, considered in the long term. In India, for example, annual rainfall that is less than 75 per cent of average is regarded as a drought. On this basis, 13.2 per cent of India's surface area undergoes a drought more than once every three years. Overall, India can vie with Africa as one of the driest places around: 151 of its districts are in arid or semi-arid regions, and they cover 54 per cent of the country and have to support 40 per cent of its population.[5] And, as in Africa, there is a widespread fear that the climate is changing for the drier.

The facts do not bear out the theory. As one investigator of the drought in Rayalaseema puts it:

> Irrigation has left us with a popular perception that this drought is more severe and more permanent than any past drought. Climate change is a myth brought on by the novelty of exponential growth in water usage . . . the falling water table is evidence of overuse of water, not of climatic change.[6]

Man-made Climatic Change

There is now no doubt that the degradation of the natural environment has had far-reaching effects on water supplies (see Chapter 4). But to add insult to injury, it is possible that these changes are also affecting climate itself, thus producing a vicious circle of cause and effect. The Canadian climatologist Kenneth Hare reported in 1984 that Africa's droughts were probably part of a natural fluctuation, even though "it is conceivable – though still unlikely in the view of some professionals – that human interference may be prolonging and intensifying the dry spells natural to the climate".[7]

Three separate mechanisms may be involved. First, where soil is stripped bare as a result of overgrazing, overcultivation and deforestation, the reflectivity of the earth's surface is increased. And where more sunlight is reflected back into the atmosphere, the atmosphere warms up, clouds are dispersed and rains become less frequent. Deserts, in other words, tend to make themselves worse.

Secondly, areas that were once vegetated may make themselves drier for another reason. Most of the rain that falls over forests, for example, comes from water that is evaporated from the leaves of the trees and undergrowth. If those trees and the undergrowth are removed, less water is available to form rain clouds. The climate gets drier. In India a study of vegetation and rainfall records over the past 100 years has shown that deforestation tends to be accompanied by both lower rainfall and fewer rainy days.[8] India's Kabini project, for example, created a 2500 hectare reservoir but also led to the deforestation of 12,500 hectares to provide space in which displaced villagers could be resettled. As a result, rainfall dropped 25 per cent, from 1500 to 1125 millimetres a year.[9]

Thirdly, there is some evidence that increasing the amount of dust in the atmosphere also decreases rainfall. In effect, more dust has the same effect as increasing reflectivity: more of the sun's energy is used to heat the atmosphere (rather than the earth itself), clouds dissipate and rains become less frequent.

No one knows how serious these man-made changes are likely to prove. More research is required before definitive conclusions can be drawn. But in regions where water supply is limited, they must now be taken seriously. Furthermore, they cannot be considered in isolation from what now threatens to become an issue of global climatic importance: greenhouse heating of the atmosphere which will almost certainly change not only average temperatures, but also rainfall, storm patterns and sea levels.

The Greenhouse Effect

Over the past century or so, the average temperature of the earth's surface has increased by approximately 0.5°C. The cause has almost certainly been a 30 per cent increase in the carbon dioxide concentration in the atmosphere, as a result of the increased rate at which fossils fuels have been burnt.

Carbon dioxide is one of many greenhouse gases that are transparent to most of the wavelengths of solar radiation but relatively opaque to the infrared radiation emitted by the earth. As levels of greenhouses gases in the atmosphere build up, a higher than normal fraction of this infrared radiation gets trapped in the earth's atmosphere, heating it up. The glass in a greenhouse has similar optical characteristics, which is why greenhouses are used to grow plants and crops that require higher than average temperatures.

There are at least 20 known greenhouse gases, and the list is growing. They include the chlorofluorocarbons used in aerosols and in refrigerants, and the nitrogen oxides emitted by car exhausts. Of these gases, carbon dioxide is not the most potent contributor to greenhouse heating, though it has been emitted into the atmosphere in far larger quantities than the other greenhouse gases. Increased carbon dioxide levels are thought to have caused about half of the total greenhouse heating so far.

Current predictions are that effective carbon dioxide levels[10] in the atmosphere will double from their pre-industrial values (about 280 parts of carbon dioxide per million parts of air by volume) some time in the middle of the next century. By 1980, carbon dioxide values had already reached 340 ppm. Without effective action to control greenhouse heating, it is expected that the earth's average surface temperature will increase by about 0.3°C a decade, providing a rise over the present value of about 1°C by the year 2025 and of about 3°C by the end of the next century.[11]

Although this figure is small, its effects will not be. The earth's average surface temperature during the last Ice Age was only about 5°C lower than it is now.

The most obvious effect will be that warmer temperatures will increase the rate at which water is used, notably in agriculture. Warmer global temperatures will have radical effects on the water cycle. Evaporation rates, and therefore precipitation rates, will increase by an estimated 7 to 15 per cent. But whether the extra rainfall will occur in the places that might benefit most is another matter. There are some indications that the greenhouse effect will be more pronounced in temperate regions than in tropical ones, with the semi-tropical regions extending polewards in both hemispheres.

At least one computer model predicts, to summarize the results in highly simplified terms, that the dry areas will get drier and the wet ones wetter. This would increase the risk of drought in dry areas and increase the risk of floods in wet ones. Such an effect might have serious negative consequences for water-scarce countries. In much of Africa, for example, soil moisture levels would fall, leading to even greater shortages in arid and semi-arid areas; all kinds of biomass would grow less well, including grasses, crops and timber.[12]

Nearly all the arid and semi-arid areas of Africa are already short of fuelwood, the fuel that provides approximately 90 per cent of rural energy needs in these regions. Any further decline as a result of increased aridity would make life impossible for many rural dwellers. Millions would be turned into environmental refugees, forced to flee their homes in search of more forested areas in other parts of Africa. The droughts of the 1970s produced just such a reaction, forcing nearly a million environmental refugees to flee Burkina Faso, and half a million more to leave Mali.

What might happen if rainfall increased in dry areas, as

other computer models suggest for regions such as India? The increased rainfall would probably occur in the monsoon period, thus intensifying the disparity between dry and wet seasons. This would not necessarily be advantageous. Unless land-use practices were greatly improved, the additional rainfall would simply swell already excessive run-off rates during the monsoon period, causing increased flooding and sedimentation (see Chapter 4).

If, on the other hand, the greenhouse effect decreases rainfall in arid and semi-arid areas, serious droughts will occur. Either way, it seems, much of the world is in for a bad time. Whatever happens, this conclusion is unavoidable since nearly a third of the world's water is currently controlled in some way or another, by dams, reservoirs, locks and sluices. The work involved has been enormous, and some of the structures erected have been the largest engineering works ever conducted. Yet all have been designed to cope with existing levels of water supply, and their anticipated variations, which have been accurately recorded over many decades. The greenhouse effect will make nonsense of many of these calculations. Adjusting the world's flood control equipment to the new order will be expensive; deciding what levels of flow to adjust it to may turn out to be largely a matter of guesswork.

The implications of greenhouse warming for those concerned with water scarcity – and many other issues – are profound. After reviewing the latest evidence at a meeting in Villach, Austria, a group of international scientists from 29 countries warned that:

> Many important economic and social decisions are being made today on . . . major water resource management activities such as irrigation and hydropower; drought relief; agricultural land use; structural designs and coastal engineering projects; and energy planning – all based on the assumption that past climatic data, without modification, are a reliable guide to the future. This is no longer a good assumption.[13]

Few detailed studies have been made of the possible effects of the greenhouse effect on water scarcity in developing countries. But US studies indicate the severity of the effects that can be expected. For example, Peter Gleick has shown that a 2°C warming could lower run-off in California's Sacramento River basin by 22 per cent; a 4°C warming would lower run-off by as much as 62 per cent.[14]

A related but equally serious effect likely to be caused by global warming is a rise in sea level caused by expansion of the warmer water in the world's oceans. Current predictions are that without action to reduce greenhouse warming sea levels will rise by about six centimetres a decade, with a predicted rise of 20 centimetres by 2030 and 65 centimetres by the end of the next century.[15]

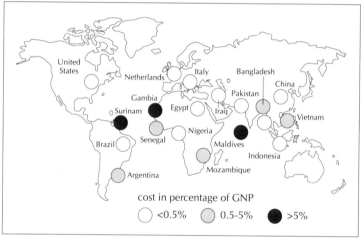

Cost of protection against a 1-metre sea level rise

Some developed countries, for example, have already begun to prepare for sea level rises. Just north of Sydney in Australia, the Warringah Shire Council has adopted new planning regulations that require developers to consider greenhouse flooding in their plans for development of the foreshore. Ground floors are required to be higher, and there are restrictions on land use for low-lying areas which mean that sites once considered suitable for hotels or luxury homes can now be used only as golf courses or car parks.[16]

A rise of 50 centimetres would cause widespread flooding in many coastal areas. This would be particularly alarming for developing countries that do not have the resources to erect expensive sea defences and many of which, such as Bangladesh, Egypt and India, have particularly high population densities in coastal areas. It is estimated that two-thirds of cultivated land in Bangladesh, for example, is already subject to flooding.[17]

A rise of 1 metre in sea level would be even more serious: at least 300 million people would be made homeless. According to the Chairman of the Coastal Impact Group of the Intergovernmental Panel on Climate Change (IPCC), Pier Vellinga, it could cost at least $13.3 billion a year to protect life and property from a sea level rise of this magnitude.[18] Many poorer nations would not be able to afford sea defences, and – in the IPCC's view – might be better advised to spend their money on adaptation strategies, such as building mounds to serve as sanctuaries during floods, than on sea defences. It has been calculated that a 1 metre rise in sea level would cover much of Egypt's and Bangladesh's arable land with sea water, flooding the ground occupied by 8 and 10 million people respectively.[19]

If current predictions are correct, by the end of the next century a third of Bangladesh and a quarter of Egypt's habitable land will be under water, as will 300 Pacific atolls. The Maldives, which is basically an atoll community, is likely to be particularly badly hit. Most of the 200,000 islanders live at a height of only 1 to 1.5 metres above sea level. The highest point in the country is 3.5 metres above sea level. And flooding swamped many of the smaller islands in 1987.[20]

While rising sea levels may make only small differences at first, the effects of tropical cyclones and storms will be greatly magnified. India and Bangladesh are particularly vulnerable to the effects of tropical cyclones, of which there have been more than eight since 1970 which have killed more than 10,000 people. Late in 1970 one storm alone killed an estimated

300,000 people in Bangladesh, and swept flood waters 150 kilometres inland.

Furthermore, a rising sea level could greatly increase salt intrusion into coastal aquifers currently used to supply both drinking and irrigation water. Many coastal areas are already fighting a battle against rising salinity, as they are forced to pump up more and more ground water to meet water needs that cannot be met from run-off alone.

Signs of a Disturbed Climate

It is too early to say with any confidence that the greenhouse effect has already started. But there are many indications that at least some of the world's climates are in a disturbed state.

Indonesia, for example, experienced its worst drought for many years during 1982–83. In the forest of East Kalimantan, the drought was of the severity expected only once in a hundred or even several hundred years. Although the area normally receives 2500 millimetres of rain a year, drought and forest fires damaged more than 3.5 million hectares of tropical forest. Drought also affected the eastern half of Australia, with lowest ever recorded rainfall over large areas. Brazil suffered from both drought and floods during 1979–83. Drought affected 37 per cent of India during 1982–83, as it did much of southern Africa, and the following year east Africa. Drought in the Sahel continued more or less uninterrupted from 1969 to 1985.

In Europe conditions varied widely. England and Wales had 214 per cent of their normal rainfall in June 1982, and this was followed by an exceptionally mild winter. There was drought in Spain, Portugal and along the Mediterranean coast. The summer of 1983 was exceptionally hot, with record temperatures recorded in England, West Germany, Austria and Switzerland. England had its second driest summer this century. In Spain the situation was catastrophic:

By mid-summer some areas of Spain had experienced three years of drought conditions. Worst hit was an area stretching from Extramadura on the Portuguese border to Valencia and the Balearics in the east, and from Toledo in the centre up to the Pyrenees. About 1.3 million people were affected by the drought. 174 towns and villages, containing 357,000 people, had little or no water.[21]

A severe drought affected many areas of Europe during 1989–90, causing a dramatic lowering of water tables in places such as south-west France, where lakes shrank to a fraction of their former size. Reduced run-off in France played havoc with that country's ambitious nuclear power programme, and electricity production had to be cut back because of lack of cooling water.

California experienced its worst two-year drought since 1906 during 1976 and 1977. Run-off dropped to half its normal level in 1976, and to only a quarter its normal level in the following year. Hydroelectric production was cut by two-thirds and more fossil fuels had to be burned, costing the State some $500 million (and doubtless adding marginally to the greenhouse effect). In some areas the water table fell by six metres, adding an estimated $25 million to pumping costs for irrigation.[22]

But what do these anomalies add up to? Amateur opinion is hardly to be trusted. However, the UN-backed Intergovernmental Panel on Climatic Change reported in November 1990 that an analysis of world weather over the 1980s showed rapid and unprecedented rates of change, with evidence of rapid global warming, melting glaciers and a decrease in snow cover. The oceans have apparently been heating up for longer, since the early 1970s, with very high temperatures recorded in the North Pacific and the tropical Atlantic, the latter probably being related to the drought in Africa during the 1980s.

The evidence that these and many similar events are signs of climatic change caused by greenhouse heating is hardly any longer even debatable. If they are a harbinger of things to come, the future looks ominous. But even if they were not, and proved to be merely the result of random variations in climate that have always been with us, they should give us pause. The world is no longer well equipped to deal with climatic variation on this scale. Too many people already live on the brink of disaster to cope adequately with even a normal share of floods and droughts, the reduction of their water supplies and the incursion of salt into their drinking water.

And it is this predicament, as we shall see in the next chapter, that is forcing the rapid deterioration of land and water resources all over the developing world.

Notes

1 J. Howell, N. Chisholm and A. Barclay, "The Background: Climate, Demography and Policy", in *Agricultural Development in Drought-Prone Africa* (London: Overseas Development Institute, 1986).

2 H. Daniel, *Man and Climatic Variability* (Geneva: World Meteorological Organization, 1980).

3 Quoted in Lloyd Timberlake, *Africa in Crisis* (London: Earthscan, 1985, p. 20).

4 WMO, *The Global Climate System: autumn 1984 to spring 1986* (Geneva: WMO, 1987).

5 Jayanta Bandyopadhyay, "Riskful Confusion of Drought and Man-Induced Water Scarcity", *Ambio*, vol. 18, no. 5, 1989.

6 K.W. Olsen, "Man-made drought in Rayalaseema", *Economic and Political Weekly*, XXII, March 14, 441–443.

7 K. Hare, "Recent Climate Experience in Arid and Semi-arid Lands", *Desertification Control Bulletin*. No. 10, May 1984.

8 Jayanta Bandyopadhyay, "Riskful Confusion of Drought and Man-Induced Water Scarcity". Op. cit.

9 N.D. Jayal, "Destruction of Water Resources – the most critical ecological crisis of East Asia", *Ambio*, vol. 14, no. 2, 1985.

10 The effective carbon dioxide level is a measure that takes account of the effects of all the greenhouse gases. If the level doubles, the effect is the same as if the level of carbon dioxide had itself doubled.

11 WMO/UNEP, *Climate Change: the IPCC Scientific Assessment* Cambridge: Cambridge University Press, 1990, published for the

Intergovernmental Panel on Climate Change, eds. J.T. Houghton, G.J. Jenkins, and J.J. Ephraums).

12 Haile Lul Tebicke, "Some Questions Towards Preparing for Climate Change in the Energy Sector in Africa". Paper delivered at the Cairo World Conference on Climate Change, 17–21 December 1989.

13 World Meteorological Organization, *A Report of the International Conference on the Assessment of the Role of Carbon Dioxide and other Greenhouse Gases in Climate Variations and Associated Impacts* (Geneva: WMO/ICSU/UNEP, 1986).

14 Peter H. Gleick, "Regional Hydrologic Consequences of Increases in Atmospheric CO_2", *Climatic Change*, vol. 10, 1987.

15 WMO/UNEP, *Climate Change: the IPCC Scientific Assessment*. Op. cit.

16 Bill O'Neil, "Cities Against the Seas", *New Scientist*, 3 February 1990.

17 Nurul Islam, *Development Strategy of Bangladesh* (Oxford: Pergamon Press, 1978).

18 Jan Sinclair, "Rising Sea Levels Could Affect 300 Million", *New Scientist*, 20 January 1990.

19 J. Broadus et al, "Rising Sea Level and Damming of Rivers: possible effects in Egypt and Bangladesh" in J. Titus (ed.). *Effects of Changes of Stratospheric Ozone and Global Climate: Volume 4, Sea Level Rise* (Washington, DC: US Environmental Protection Agency and UNEP, 1986).

20 Sue Wells and Alasdair Edwards, "Gone with the Waves", *New Scientist*, 11 November 1989.

21 The quotation and other figures in this section are from Michael Glantz, Richard Katz and Maria Krenz (eds.), *The Societal Impacts Associated with the 1982–83 Worldwide Climate Anomalies* (Boulder, Colorado: National Center for Atmospheric Research and UNEP, 1987).

22 Peter H. Gleick, "Climate Change and California: past, present and future vulnerabilities", in M.H. Glantz (ed.), *Societal Responses to Regional Climate Change: forecasting by analogy* (Boulder, Colorado: Westview Press, 1988).

Chapter 4

The Degradation of Land and Water

"Forgive us Aral. Please come back."

Slogan written on fishing boat lying on
the dried up bed of the Aral Sea

The climate-related problems described in Chapter 3 are now being worsened almost everywhere by human action – by the destruction and degradation of both land and water resources. As multinationals fell forested hillsides in search of valuable hardwoods, they leave fragile soils exposed to harsh monsoon rainfalls. The soils are washed away. As subsistence farmers are forced further on to marginal lands, they remove the vegetation that protects the soil from fierce erosive winds. The soil then blows away. As pastoralists graze more and more stock near major waterholes, the soil gets trampled down and compacted.

All this affects more than just food production because surface vegetation and soil act like a giant sponge, playing a critical role in the regulation of water flow and storage. This sponge stores excessive rainfall when it occurs, releasing it gently into streams, rivers and lakes during drier periods when water is most needed. If land is cleared of its vegetation and soil, its ability to absorb and store water is reduced. This means not only that soil becomes less fertile, but also that a larger amount of rainfall runs directly into rivers, affecting river levels and water turbulence, and the amount of sediment carried in the water.

A study in Tanzania showed that in areas with a natural

cover of grass, bush, and scattered trees, no erosion occurred and almost all rainwater was retained in the soil. On cultivated land, 25 per cent of the rainwater ran off; and completely uncovered land lost 50 per cent of rainwater.[1]

Deforestation can, therefore, significantly increase flow of water in rivers: cutting down pines in Colorado has been shown to increase stream flow by about 30 per cent while removing all woody vegetation from a watershed in North Carolina increased stream flow by more than 70 per cent during the first year.[2] Even more dramatic consequences were noted in the Australian Alps, after a fire in the early 1970s destroyed 265 square kilometres of forest. Rainstorms which usually produced peak flows of 60–80 cubic metres were producing peaks of 370 cubic metres – over four times higher – after the fire.[3]

Conversely, experiments in the United Kingdom have shown that planting forest cover in a denuded watershed can reduce the total flow of water from the watershed by as much as 40 per cent. Similar results were found in the United States, when reforestation was carried out in an area of Tennessee so completely eroded that agriculture had been abandoned there. Within five years, the volume of river water had dropped by 50 per cent, and peak flows had been reduced by 90 per cent.

When soil disappears, the natural cycle of water regulation breaks down. Since there is nothing to store water, heavy rains run straight off the barren surface, gouging out deep furrows or gullies as they go. Gullies like these can be seen in many of the world's deforested areas but particularly in a band stretching from Ethiopia, through central Africa, and down to Lesotho in the south. These gullies, now often tens of metres deep and kilometres long, were once cropland. They are the symptoms of a ravaged land.

The vicious circle of damage does not stop with the destruction of cropland. Soil from what was once fertile land is washed away by heavy rain when vegetation is removed. It ends up in places where it is not wanted, such as rivers, reservoirs and lakes. As a result, rivers turn to shallow, muddy torrents, supporting neither fish nor navigation. Flooding increases.

Waterholes dry up. In the end, green and pleasant lands turn to brown and arid ones.

The Wettest Desert on Earth

The Cheerapunji region of India, for example, has one of the highest recorded rainfalls in the world – more than nine metres a year, all of which falls during the monsoon. But reckless deforestation and the unlawful felling of trees which have been going on for 25 years have all but removed the region's ability to retain water. As the trees have gone, the thin layer of soil in which they grew has been washed away by the heavy rains. These rains now run straight down the hillsides into neighbouring Bangladesh, where they cause severe flooding.

Within two months of the monsoon, Cheeranpunji's enormous rainfall has gone, even more quickly than it came. Thus one of the wettest regions of the world is now suffering from water shortage and, in some areas, dessication.

Similar problems are now threatening the Panama Canal[4] where every year 12,000 ships pass between the world's two major oceans – bringing Panama, incidentally, its main source of foreign exchange, equivalent to some $350 million annually.

But now the canal is drying up, threatening its own survival. Because of the system of locks engineered into the canal, every ship needs about 200 million litres of water to pass through the canal. The water for this operation is supplied by the heavy rains that fall on the country's mountain ranges. The rainwater is collected in the Gatún Lake that was created by damming by John Stephens, the engineer who built the canal. The canal is fed from this lake, and water flows from it, out towards the two oceans, to fill up the locks every time they are opened to let a vessel pass through. In 1935 a second lake was created, the Madden Lake, which also provides water and hydroelectric power for Panama City and Colón.

Until the late 1970s this system worked well. But then decades of indiscriminate forest felling in the mountains showed their effects for the first time. There was a serious drop in the canal's water level. Panamanians were faced with a difficult problem: either to close the canal to ships, or divert more water from the Madden Lake to the canal and deprive local populations of the water and power they needed. The canal was closed, forcing shipowners to make a three-week detour round the Cape of Good Hope. The same thing happened in the dry periods of 1981 and 1982.

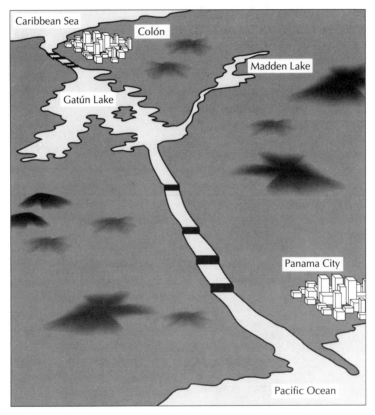

Panama Canal

The problem is caused by the loss of thousands of hectares of forest every year from the surrounding areas. This is removing the sponge-like effect of the forest floor, destroying the mountains' ability to store the heavy rains that fall during the wet seasons. Much of this water was previously released slowly into the two lakes during the dry periods.

As the forest is removed, its soil is washed downstream into the lakes below. The sediment that is deposited reduces the storage capacity of the lakes and lowers the depth of the canal. The Madden Lake already holds 5 per cent less than it did in 1935, and the rate of siltation is increasing. The canal needs repeated dredging as its capacity shrinks.

This adds to an already acute problem, in that the canal never was large enough to handle supertankers. Proposals for a new sea level canal that would not need freshwater feeding, and could use ocean water, are now being examined.

Meanwhile, political controversy surrounds the issue. The main reason for forest felling in Panama is that most of the good agricultural land is owned by a few rich farmers. The *campesinos* are forced either to join the urban poor, or take to the mountains where they survive by slash and burn agriculture, or by clearing the forest to raise cattle. The government has responded by making it illegal to cut any tree more than five years old – a law it is unable to enforce. USAID created a fund to support the moratorium on logging, but the US withdrew this and all other forms of humanitarian aid in the period of political unrest before Noriega's surrender to US forces. Panama retaliated by throwing out USAID.

The solution, of course, is to find other employment or other land for the campesinos. It is this basic problem that is threatening Panama's forest, and hence its water supplies and canal.

How Droughts Lead to Floods

The sediments that were once soil now clog many of the world's major rivers. The floods that plague Bangladesh on a yearly basis have now become infamous – and are likely to become more so if sea levels rise as a result of greenhouse warming. Though hard to prove, it is likely that soil erosion and deforestation have already taken a savage toll in the destruction of lives and property in many tropical countries. Certainly in both South America and South Asia there is widespread acceptance that floods and mudslides have become both more frequent and more serious. A 1984 study by Earthscan and the Swedish Red Cross reported that floods affected 5.2 million people per year in the 1960s and 15.4 million per year in the 1970s.[5]

Bangladesh has large areas which have suffered deforestation and, as a result, the country experiences both severe flooding and, subsequently, drought. Roy Ward, a British geographer, said of Bangladesh:

> It is ironical that in an area through which flows the discharge from the second largest river system in the world, flood is followed by drought because the floods, which assume such vast proportions, often inundating one-third of the total land area of Bangladesh, are not adequately controlled. Destruction by flood is thus followed by drought-provoked famine in a vicious and, as yet, unending circle.[6]

This pattern is repeated throughout much of the Indian subcontinent. As the slopes of the Himalayas are cleared for habitation and agriculture, flooding is increasing throughout northern India, much of Pakistan, Myanmar and Bangladesh. The death rate from floods in India in the 1980s was 14 times greater than from those in the 1950s. According to the country's National Commission on Floods, the area annually afflicted by floods has grown from 25 to 40 million hectares in three decades.[7]

In 1978, India suffered a bout of severe flooding after only two days of particularly heavy rainfall. More than 2000 people drowned, 44,000 cattle were lost and 66,000 villages were inundated. This, of course, is by no means the most serious flood on record. One contender – excluding the Biblical flood – is the flood caused by a tropical storm in the Bay of

Bengal during 12–13 November 1970 which killed 225,000 people.

The loess plateau of central China is also the cause of severe flooding. Loess is a fine yellow silt that covers more than a quarter of a million square kilometres of China's heartland. Heavy seasonal rains are sweeping some 10,000 tonnes of it off every square kilometre of the plateau every year. Flushed into the Yellow River, the silt turns the river into a swirling yellow torrent. Some 2000 million tonnes of loess are thus carried annually in the river, three times as much sediment as can be found in any other river.

By the time the Yellow River reaches Kaifeng, still 800 kilometres from the sea, the river may be 40 per cent silt. This is continuously deposited on the riverbed, which is rising by up to a metre a year. Dykes along the side of the river therefore have to be built up at the same rate, and for much of its way over the Yellow Plain to the sea the river is eight metres above the surrounding fields.[8]

Floods along the banks of the river are annual events, leaving behind them a trail of destruction of both lives and property.[9] The river – known as "China's Shame" – has killed more people than any other single feature of the earth's surface.[10] In 1887, the Yellow River flooded, killing between 900,000 and two million people by drowning and starvation; and in 1931 one of the world's worst floods left between one million and 3.7 million people dead.

Similar stories abound in Latin America, where both floods and mudslides make regular news. In 1983, floods and landslides in Ecuador killed scores and left $400 million worth of crop and property damage; Bolivia suffered record floods; and in Peru rains washed away whole villages and swept buses off mountain highways near Lima. The Peruvian government declared a state of emergency as disease broke out in flood-ravaged areas.

Of all the natural disasters, floods cause the most deaths, accounting for 40 per cent of all deaths from natural disasters (twice as many as the next highest category, tropical cyclones). It is no surprise that 95 per cent of these deaths occur in developing countries, and only 5 per cent in the industrialized world.[11]

In many of the large cities in developing countries, the poorest people live in buildings and areas unregulated by government controls. These dangerous sites are often the only areas left for the urban poor to settle in, and many are prone to flooding and landslides. In Guayaquil, Ecuador, it was estimated in 1975 that 60 per cent of the population of over one million lived in squatter communities built over tidal swampland. The small, insubstantial houses are built on poles above water prone to flooding and storm surges. For some people, dry land is a 40-minute walk away over timber catwalks. Conditions are as dangerous in Mexico City, where some 1.5 million people live on the drained lake bed of Texcoco, which floods or becomes a bog when it rains.

As riverside towns and cities expand, it is usually the poorest people who move down to live in the areas prone to flooding. Delhi has expanded on to the flood plain of the Yamuna River, and it is estimated that over 1.5 million people there are at risk from floods. In 1980 the Yamuna River flooded to form a lake of some 130 square kilometres. Floods in Bangladesh kill hundreds of thousands of people partly because 15 per cent of the 90 million inhabitants live less than 3 metres above sea level. However, flood plains remain desirable places to live, as land is fertile and transport easy and, even knowing the risks, people often return to the same area when floods subside.

The economic costs of floods are also extraordinary – and apparently mounting. Direct damage was estimated at more than $1750 million annually in the early 1970s, and must be much greater now. But indirect damage, just in terms of the shortened lives of so many hydroelectric and storage reservoirs, must be even greater. Few reservoirs in developing countries are now expected to have a useful life of more than 30 years – and many have served their purpose for only 15 or 20 years, though originally planned to last much longer (see Chapter 8).

Irrigation and the Environment

Every year about 3300 cubic kilometres of water are used to provide water for the one-third of the world's harvest that now comes from irrigated land. In all, 17 per cent of the world's cropland is irrigated, and productivity over much of it is high.

In countries such as Egypt, all cropland is irrigated, since rainfall is simply too low to permit rain-fed agriculture. And in some of the largest countries in the world, irrigation accounts for the bulk of production: one-third of Indian cropland, and nearly half of Chinese cropland, together accounting for more than 100 million hectares, are irrigated. Without irrigation, food production over most of the arid and semi-arid tropics would be impossible. Many countries would lose more than half their domestic food production.

Ever since the decline of the first Mesopotamian civilizations, irrigation has been associated with environmental problems that, throughout history, have often proved catastrophic – both for the cropland being irrigated and for the sources used to provide the irrigation water. There are three main dangers: waterlogging, salinization and the depletion of underground aquifers.

Waterlogging occurs either as a result of the repeated downward percolation of irrigation water or when repeated irrigation raises the level of the water table to such an extent that the roots of crops become literally submerged, killing them or reducing production dramatically. Water tables under irrigated land can rise quite rapidly; on one project in India, the water table rose by an average of one metre a year over a five-year period. On another project in Uttar Pradesh, irrigation was extended to cover a further 400,000 hectares of cropland; at the same time some 500,000 hectares of irrigated land had to be taken out of production due to waterlogging – a net loss to the project of some 100,000 hectares.[12]

Waterlogged soil is virtually useless, and solutions are expensive. They include lining irrigation canals to stop leakage, removing ground water by pumping, and removing surface and ground water through better drains (this is the "plug, pump and drain" approach of water engineers). Educating farmers to use irrigation water more sparingly, and to maintain irrigation systems better, can be an even longer job. In spite of the obvious dangers of waterlogging, irrigation systems were still being installed without adequate drainage until long after World War II.

Curing waterlogged land can be impossible if the rising water levels are also contaminated by salts or other chemicals. Salinization is caused by the continual evaporation of irrigation water in fields where the salts that all water contains become increasingly concentrated. Eventually, salt levels can become so high that crops die and the soil is effectively poisoned. Whitened fields are a common sight for air travellers these days, and salinization has taken a massive toll of the world's most productive croplands over the past few decades. The problem has become so bad in parts of the Soviet Union, for example, that thousands of tonnes of soil have had to be physically removed from contaminated sites, and replaced with new soil from other areas.

Although salinization can be caused by the use of contaminated water, it can also occur after prolonged irrigation with relatively clean water. Even the purest water contains 200–500 parts per million (ppm) of salts (compared to the seas' 35,000 ppm). If, as if often the case with irrigation, a metre of water is used over the course of a year, each hectare of land potentially receives several tonnes of salts. With good irrigation practice, adequate drainage, and frequent "rinsing" of the soil, this does not matter. But without good management it does. If these salts are allowed to accumulate, they will – in the end – turn the land underneath into abandoned farmland.

Probably as much as one-quarter of the world's irrigated land has already been damaged in this way.[13] Certainly salinization affects productivity on at least 38 million hectares among the world's top five irrigating nations: India, China, the United States, Pakistan and the Soviet Union. In India, more than 7 million hectares of salinized land have already been abandoned.

Irrigation can be provided in one of two basic ways: either rivers and streams can be dammed to form large reservoirs from which irrigation water can be drawn as needed; or ground water can be pumped from beneath the soil surface and piped to the fields, perhaps some distance away, where irrigation is required.

Large dam projects are currently heavily criticized for both environmental and humanitarian reasons. As a high tech solution to Third World water problems, they seem beset by problems (see Chapter 8). However, these problems relate mainly to their faulty execution and planning. There is, of course, no intrinsic reason why large dam projects should not provide humanity with net gains.

Groundwater pumping may be intrinsically more dangerous. For one thing, irrigated agriculture can be regarded as sustainable only if ground water is used no faster than it is replenished. All too often, it is extracted faster – frequently much faster – than its natural replenishment rate. One result is that pumping becomes more costly, and irrigation more expensive. But others who also depend on the aquifer being pumped may also suffer. Village tubewells can dry up, streams run dry and lakes disappear. To make matters worse, these effects occur mainly in the region whence the water is being pumped, rather than where the irrigation is needed.

While this problem is common in developing countries, it occurs in industrialized areas as well. In the United States, for example, it is estimated that one-fifth of irrigated land is fed from ground water faster than the replenishment rate.[14] By the early 1980s, water tables were falling by as much as 15 centimetres a year under irrigated land in much of Texas, and this was one reason why the area of irrigated land in the state fell by 30 per cent between 1974 and 1987.

A similar situation exists in many other countries, though the statistics are less easy to come by. Overpumping in India's Tamil Nadu state, for example, has been reported to have lowered the water table by 25–30 metres in a decade. Else-where in India, pumping in coastal regions has had other unfortunate effects, including the invasion of salt water into underground aquifers previously used to supply drinking and irrigation water.

The Shrinking Aral Sea

The world's fourth largest freshwater lake – the Aral Sea in the southern Soviet Union – has now been virtually destroyed by irrigation. Two rivers feed – or fed – the sea in this semi-arid area where nearly all agriculture relies on irrigation. The region supplies much of the Soviet Union's cotton, fruit, vegetables and rice. But expanding the area under irrigation – now totalling 7.5 million hectares – to keep pace with Soviet demand for food and fibre has meant diverting more and more water from the rivers that supply the Aral Sea.

Over the past 30 years, two-thirds of the water that was once in the sea have drained away. Salinity has tripled, turning much of the coastline into a kind of swampy saline bog. Since 1960, the water level has fallen by no less than 13 metres, and the sea's surface area been halved. Extrapolations to the year 2000 suggest that by then the level will have dropped another 10 metres or so, and salinity levels tripled again.[15]

There is no controversy about the cause. So much irrigation water is used in the semi-arid areas to the south and east that the rivers that fed the Aral Sea are almost expended before they even reach it. The Syr Darya never reached the sea at all between 1974 and 1986. The Amu Darya failed five times between 1982 and 1989, mainly because so much of its water is now diverted into the Kara Kum irrigation canal 800 kilometres away. This canal, the longest in the world (more than 1300 kilometres), delivers water from the Amu Darya to Turkmenia, along Iran's borders.

As the Sea has dried up, so has the land become progressively poisoned by attempts to grow irrigated crops over ever larger

The shrinking Aral Sea

1960

1973

U.S.S.R

Aral Sea

1989

2000?

areas. As Vasily Selyunin, one of the Soviet Union's most respected writers, puts it:

> The root of the problem is over-irrigation on a scale so vast that it has washed all the humus out of the soil. The loss had to be made good with shock doses of chemical fertilizer. As a result, the earth has become like a junkie, unable to function without its fix . . . In the Soviet Union as a whole, two kilos of chemicals are spread on each hectare of ploughed land. In Central Asia it is more like 50 kilos.[16]

Already much of the area has been reduced to economic ruin. Some of what were previously fishing ports now lie more than 40 kilometres from the lake's shore. The fish catch, once 25,000 tonnes a year, has been reduced to zero. In the city of Aralsk, half the population has fled. Quays that served to unload the laden fishing boats of the past now jut out into empty desert. Dramatic pictures of stranded fishing boats illustrate the grim reality of the dry statistics of decline. As a *Sunday Times* writer puts it:

> Once it was a fruitful sea. Now it is a poisoned wasteland. In the ships' graveyard of the Aral, the Muinak fishing fleet lies where it fell when the water drained away, waiting to be dismembered – by former fishermen.[17]

Were things allowed to continue in this way, the Aral Sea would inevitably be reduced to warm puddles of water far more saline than the oceans themselves. To stop this happening, a long-term plan, announced in 1988, has been made to increase flow into the Sea by a factor of three over the next 20 years by making the use of agricultural water more efficient. The plan is expected to cost at least $50 million.[18] Costly though this may be, the alternative, of taking land out of production in one of the Soviet Union's most productive agricultural areas, is hardly appropriate to a nation that imports substantial amounts of food.

The conflict between food and water appears in stark relief in the republics bordering the Aral Sea. And outside the Soviet

Union's Ministry of Water Resources, few believe that the plan to save the Aral Sea has any chance of success. It seems unlikely now that the 600 cubic kilometres of water that the Sea has lost can ever be put back. Critics of the plan therefore conclude that the Sea has gone forever.

The official solution is to increase the level of water flowing into the Sea so that it reaches some 20 cubic kilometres a year by the year 2005. Comments Selyunin:

> I can tell you precisely what will happen to the Aral Sea at that rate. By 2005 it will be reduced to 17 per cent of its original volume, and will be divided into three medium-sized lakes. What is a promise to "save" the Aral Sea? It is a lie.[19]

The last word goes to the young Uzbek poet Mukhammed Salikh. "You cannot fill the Aral with tears", he writes.[20]

Doom in the Doon Valley

Elsewhere there is conflict between water and minerals. Heavy quarrying in water-bearing strata, for instance, can reduce important water sources to a mere trickle.

The Doon Valley in the Indian foothills of the western Himalayas is a case in point. Although rainfall is plentiful, amounting to an average 2000 millimetres a year, water supplies there have been drastically reduced as a result of quarrying for limestone at the northern end of the valley in the Mussoorie Hills. This limestone has, over millennia, collected and stored much of the region's copious rainfall. Over the past 25 years, however, the limestone has been extensively quarried for industrial use. As a direct result, stream flow in the Doon Valley has declined 60 per cent, and most of the perennial rivers that were fed by the limestone aquifer now carry only monsoon flood water. Both crop and livestock production have declined by 50 per cent, and what was once productive farmland is going out of production.

Villagers were so threatened by these developments that

they appealed to the High Court of India, as a result of which limestone quarrying in the valley has now been strictly controlled.[21] But elsewhere in India iron ore mining is adding huge volumes of sediment to the Tungabhadra River in Western Ghat; and it has been estimated that each 10 metre stretch of road constructed in the Himalayas adds an additional 2-tonne load to the debris annually swept into the area's rivers, reservoirs and flood plains.[22]

The Spreading Deserts

The Doon Valley and the Cheerapunji regions of India are examples of the insidious process of dessication. While deserts occur mainly in areas where rainfall is low, and surface water scarce, land where water is simply unavailable for much of the year can easily become dessicated as a result of deforestation, mining and even the waterlogging and salinization caused by poor irrigation management. While dessication is the correct term to describe the process by which land in areas of adequate rainfall is deprived of moisture, the United Nations agencies have insisted on using the term desertification to describe dessication.

UN studies claim that no less than 29 per cent of the land surface now suffers from slight, moderate or severe desertification – and a further 6 per cent is "extremely severely desertified". Some 230 million of the 850 million people who lived in the world's drylands were estimated to be suffering from the effects of severe desertification in 1984. No less than 18.5 per cent of land in South America, Asia and Africa is now severely desertified.

The problem is getting worse. In spite of attempts both by individual countries and by international agencies, the relentless march of the desert cannot apparently be halted. More than 20 million hectares of land become so desertified every year that agriculture becomes uneconomic. Some 6 million hectares are permanently degraded to desert every year.

While exceptionally dry years in the 1960s to 1980s in parts of Africa have played a part in this process, desertification is broadly the result of human action and inaction – deforestation, overcropping, the use of unsuitable land for arable farming, the increase of livestock numbers on rangelands, and the pressures of cash cropping have all played their part.[23]

Spending Fast to Produce Less

As populations increase, there is inevitable pressure on natural resources. More people chase fewer natural goods. This is a recipe not for disaster but for hard thinking. How can the demands of the planet's population best be served by what the planet has to offer?

Not, surely, by overworking the land, felling the forests and mining ground water for irrigation?

In many areas, the problem is so severe that salinization is slowing down the rate at which new land can be brought into production to feed the new mouths produced by expanding populations. In the 1980s, at least, the amount of land going out of production from causes such as salinization, waterlogging and desertification was roughly equal to the amount of new land being brought under the plough for the first time. In some places, the area of irrigated land was contracting at just the time when expansion was needed. During the late 1970s and early 1980s in the Sahel, for instance, irrigation was being extended to 5000 new hectares a year. But, according to the Club de Sahel, waterlogging and salinization were taking another 5000 hectares out of production.

"Running fast to stand still' used to be the term applied to non-constructive activity of this kind. But where such action also deprives pastoralists and other small-scale farmers of the meagre living they once had, it might be better described as "spending fast to produce less".

Notes

1 A. Wijkman and L. Timberlake, *Natural disasters: acts of God or acts*

of man? (London: Earthscan,1984).

2 H. G. Wilm and E. G. Dumford, *Effect of Timber Cutting on Water Available for Stream Flow from a Lodgepole Pine Forest* (USDA Technical Bulletin, 1948).

3 A Wijkman and L. Timberlake, *Op. cit.*

4 Paul Simons, "Nobody loves a canal with no water", *New Scientist*, 7 October 1989.

5 A. Wijkman and L. Timberlake, *op. cit.*

6 A. Wijkman and L. Timberlake, *op. cit.*

7 Erik P. Eckholm, *Down to Earth: environment and human needs* (London: Pluto Press, 1982, p. 173).

8 A. Wijkman and L. Timberlake, *op. cit.*

9 Tim Grout-Smith. "Profit and Loess from China's Silt", *New Scientist*, 9 September 1989.

10 A. Wijkman and L. Timberlake, *op. cit.*

11 Adrian T. McDonald and David Kay, *Water Resources: issues and strategies* (London: Longman Scientific and Technical, 1988.)

12 Indian Ministry of Agriculture, *Report of the Task Force on Increasing Agricultural Productivity in the Command Areas of Irrigation Systems.* (Ministry of Agriculture, 1984, *The Mukerji Report*).

13 Sandra Postel, *Water for Agriculture: facing the limits* (Washington, DC: Worldwatch Institute, 1989, Worldwatch Paper 93).

14 Postel, *op. cit.*

15 G. N. Golubev, "Economic Activity, Water Resources, and the Environment", *Hydrol. Sci. J.* 28(1), 57-75

16 Vasily Selyunin, "A Hundred Pages of Anguish", *The Sunday Times*, 25 February 1990.

17 "The Stolen Sea", *The Sunday Times*, 25 February 1990.

18 Perera, Judith, "Kremlin Moves to Save the Aral Sea", *New Scientist*, 26 November 1988.

19 Selyunin. Op. cit.

20 William S. Ellis, "A Soviet Sea Lies Dying", *National Geographic*, February 1990.

21 J. Bandyopadhyay and V. Shiva, "The Conflict Over Limestone Quarrying in Doon Valley", *Environ. Conserv.*, 12, 131-139, 1985.

22 V. V. D. Naraya and Rambabu, "Estimations of Soil Erosion in India", *Journal of Irrigation and Drainage Engineering*, 109, 409-434, 1983.

23 UNEP, *General Assessment of Progress in the Implementation of the Plan of Action to Combat Desertification 1978-1984* (Nairobi: UNEP 1984).

Chapter 5

Living with Water Scarcity

Two-thirds of the African population will live in severely water-stressed countries within only a few decades.

Malin Falkenmark

Water shortage has many dimensions. Because irrigation is so much the heaviest user of water, crop production is the first area to suffer when water supplies are inadequate. In the battle for water, it is cities, not farmers, that usually win. Industry suffers next. Supplies of cooling water, and water for household needs, are usually the last to be cut back.

Defining the levels at which hardship begins and plenty ends is notoriously difficult. One of the few scientists to attempt to do so is the Swedish hydrologist Malin Falkenmark. She has defined what she calls five "water competition intervals". Countries which have 10,000 cubic metres of water per person a year or more have limited water problems; those with 1670–10,000 have general problems; those with 1000–1670 can be regarded as "water-stressed"; those with 500–1000 suffer from chronic water scarcity; and those with less than 500 cubic metres of water per person a year are beyond what Falkenmark calls the "water barrier".[1]

On this basis, the predicament faced in the 1980s by a sample of countries was as follows:

Country	Internal water availability* (m³/cap/a)	State	Withdrawal as % of available water†
Canada	109,510	limited problems	1
Panama	60,760	limited problems	1
Nicaragua	46,730	limited problems	1
USA	10,060	limited problems	19
China	2520	general problems	16
India	2270	general problems	21
Peru	1840	general problems	–
Haiti	1500	water stressed	–
S. Africa	1400	water stressed	18
Poland	1290	water stressed	30
Kenya	610	chronic water scarcity	–
Tunisia	490	beyond "water barrier"	53
Israel	370	beyond "water barrier"	88
Barbados	200	beyond "water barrier"	51
Libya	170	beyond "water barrier"	374
Malta	60	beyond "water barrier"	92
Egypt	40	beyond "water barrier"	97

* internal availability, excluding river flow from other countries
† total availability, including river inflow

Source: World Resources Institute. *World Resources 1986 – an assessment of the resource base that supports the global economy* (New York, Basic Books, 1986).

Several important points need to be made about this table. First, the figures are only approximate. Secondly, the table does not compare like with like. The figures of water availability in the left hand column include only renewable water, excluding inflows from other rivers. The final column, which estimates withdrawal as a percentage of what is available, uses a different definition: annual renewable water plus net river inflow.

Net river inflow can be a critically important parameter. Egypt is one of the driest countries in the world if you exclude from its water resources what the Nile brings it.

To do so, however, gives a totally false impression of the water-richness of Egyptian society. The Nile actually provides Egypt with 50 times more water than does rainfall. Providing, that is, that upstream countries do not interfere with Egypt's river supply. Egypt and the Sudan have a formal agreement on how much water Sudan can divert. Ethiopia, however, is not party to this agreement (see Chapter 7); since 86 per cent of Sudan's Nile water comes from Ethiopia, that country could wreak havoc to downstream water use if it so chose. International conflict over river water use is dealt with in Chapter 7.

The figures in the table, however, deal only with renewable water. This explains the anomaly of Libya which apparently uses nearly four times as much water as it has. This is because Libya relies extensively on a huge underground aquifer for its water supplies. To all intents and purposes, this water is not renewable in the sense that its recycle time is measured in thousands of years at least. Some of the water may actually be connate or fossil water which will never be renewed.

The World Resources Institute (WRI) examined the per capita availability of water in 100 countries in 1986, and came up with an assessment in which more than half of the countries included had low or very low water availability:[2]

Category	Water availability (m³ per person per year)	Countries (%)
very low	1000 or less	14
low	1000–5000	37
medium	5000–10,000	14
high	10,000 or more	35

According to the WRI a critical point is reached when a country uses more than a third of its available water. In the WRI survey, 19 per cent of the countries surveyed fell into this category in 1986. This figure will be substantially higher in the year 2000 as a result of population growth. In most countries human populations are growing while water availability is not. What is available for the use of each man, woman and child on the planet is therefore falling.

The effects of population growth are easy to demonstrate. Average annual water use is currently about 800 cubic metres per person. World population is increasing at the rate of about 80 million a year. This implies an increased demand for water of 64 billion cubic metres or 64 cubic kilometres a year, the typical flow rate of a large river roughly the size of the Rhône, the Rhine or the Euphrates. Most of this growth occurs in water-stressed countries where demand is already knocking at the door of supply.

In the coming decades, it is inevitable that population growth will push many developing nations into conditions of chronic water scarcity; others will actually cross the "water barrier". For them, plans for development will be governed – and probably limited – by issues of water supply and management. In the words of the *Global* 2000 report:

> By the year 2000 population growth . . . will cause at least a doubling in the demand for water in nearly half the countries of the world . . . Much of the increased pressure will occur in the developing countries when, if improved standards of living are to be realized, water requirements will expand several times. Unfortunately, it is precisely those countries that are least able, both financially and technically, to deal with the problem.[3]

The situation is likely to be particularly critical in Africa. On paper, Africa is relatively well endowed with water, with a total run-off considerably larger than that of Europe, for example. However, most of the continent's water drains west into the Atlantic. The Congo alone, which is the second largest river in the world, delivering 1300 cubic kilometres a year, carries away nearly one-third of the continent's run-off. There is scant prospect of being able to redistribute this water towards the water-poor regions in south and east Africa, and even less of channelling it to the very arid countries of north Africa.

This leaves three regions of Africa – the north, the east and the south – with scant supplies. Furthermore, populations in Africa are expanding faster than anywhere, with growth rates in several countries still in excess of 4 per cent a year. With populations growing at this speed, or even somewhat slower, theoretical water supplies per capita are bound to fall. As they do, the degree of water stress will alter, and many countries will be taken into the areas that Falkenmark describes as water stressed or chronically short of water. Several will even cross the "water barrier".

Falkenmark estimates that by the end of the century – now less than a decade away – 11 African countries will enter the categories of chronic water stress or lie beyond the "water barrier". They include four countries in East Africa (Burundi, Kenya, Rwanda and Somalia) and all five North African countries (Algeria, Egypt, Libya, Morocco and Tunisia). The remaining two countries, Malawi and Zimbabwe, are in southern Africa.

By then some 250 million Africans will be living in water-stressed countries, suffering from water deficiencies on a scale roughly comparable to those that exist today in the arid valley of the lower Colorado River. Of these, 150 million will be living in countries where there is chronic water scarcity. One country, Tunisia, is already beyond the "water barrier".

Twenty-five years later, by the year 2025, another 10 countries will be water stressed – Ethiopia, Tanzania, Uganda, Lesotho, Mozambique, Benin, Burkina Faso, Gambia, Nigeria and Togo. Falkenmark estimates that no fewer than 1100 million Africans will by then be living in water-stressed countries, some two-thirds of the African population. Another four countries – Kenya, Rwanda, Burundi and Malawi – will be beyond the "water barrier".[4]

While Africa provides the most dramatic example of Falkenmark's thesis, the outlook is far from optimistic else-where. Israel, for example, will have long passed the "water barrier" by the year 2000. Saudi Arabia, which has already passed it, will be relying increasingly on desalination and importing water from elsewhere. Water problems are likely throughout the Middle East, particularly in Lebanon, Jordan and Syria. India and China will both be using substantial fractions of their renewable supplies, and India may be exceeding its supply.

Even in what are normally regarded as the water-rich countries of the industrial north, there are likely to be serious problems. According to a survey prepared by the UN Economic Commission for Europe,[5] five such countries already have inadequate water to meet current needs: Cyprus, Malta, Poland, Romania and the Ukraine SSR. And by the year 2000 another five are likely to join their ranks: Bulgaria, Greece, Hungary, Luxembourg and Turkey.

On this basis, the future looks bleak indeed. But the Malthusian picture of an expanding population desperately trying to extract more and more water from a limited resource has been challenged by many scientists.[6] As Falkenmark says, "The relevant question to ask is *not* how much water do we need and where do we get it, but rather: how much water is there and how do we best benefit from it?"

Water use, like water itself, is a fluid concept, and dramatic changes in demand can be effected by small changes in both technology and custom. And, contrary to popular belief, human attitudes can change very swiftly in the face of crisis. Who would ever have guessed, for example, that the energy-dependent West could have made such dramatic reductions in energy use during the oil crisis of the 1970s? The same is certainly possible with water. Indeed, even more dramatic savings could be expected since, unlike water, energy has always been regarded as relatively expensive.

Industrial and Domestic Water Savings

How much water is used for specific industrial purposes,

for example, can vary enormously depending on the factory, the process used and the country in which it is situated. In temperate regions, at least, water has nearly always been regarded as virtually limitless and practically free. Industry has made extravagant use of it, without regard for either cost or effects on supply. But, as the cost of water has risen, and supply limitations become more widely publicized, industry has made many successful attempts to reduce its use – in some case, by large amounts. Water is being increasingly recycled, and many sources of waste eliminated. There are reports of declining industrial per capita use of water from many countries, including France, Japan, the United States and the Soviet Union.[7]

For example, water-stressed Israel, which already exploits all its reliable run-off, managed to reduce its industrial use of water from 20 cubic metres per $100 worth of production to 7.8 cubic metres between 1962 and 1975.[8] A large textile plant in São Paulo, Brazil, cut its water use by 39 per cent within two years of the city levying a charge for treating its effluent. The savings were made by turning off taps when no water was needed and using wash water more than once. Sweden has demonstrated a pulp- and paper-making technology that uses half as much water as conventional processes but produces 95 tonnes of paper from 100 tonnes of wood, compared to the 50 tonnes produced previously.[9]

There is also enormous waste of water. It is estimated, for· example, that no less than 30 per cent of the United Kingdom's public water supply leaks away before it reaches the user. Redesigning domestic appliances such as showers and toilets can produce at least a 50 per cent saving in domestic use.

Sometimes water can be saved by combining the wastes from one use with the supply of another. Indeed, this kind of sequential reuse will become critically important for all countries with impending shortages. Hillel I. Shuval, the Director of the Division of Human Environmental Studies at the Hebrew University of Jerusalem, writes:

In Israel, total waste water reuse is now the declared national policy. Some 30 per cent of all urban waste water is currently being utilized mainly in agricultural irrigation. This has resulted in an increase in available water resources and has reduced pollution of streams and coastal recreational areas. Irrigation with waste water has the additional benefit of supplying all of the nitrogen and most of the other nutrients normally required by agricultural crops.[10]

All sewage discharges to the Mediterranean coast of Israel have now been returned to the land.[11] Not only is this good news for the Mediterranean, but it also helps provide fertilizer for the land as well as saving water. However, Israel may well have reached its upper recycling limit. There is evidence that natural run-off cannot be increased by more than 25–30 per cent through the reuse of waste water. By the year 2000 Israel plans to recycle 80 per cent of its waste water.[12] Even so, Israel's water gap is thought to have already reached 200–300 million cubic metres a year – reason enough for that country's determined effort to perfect (and resuscitate) the gentle art of water harvesting to boost agricultural production (see Chapter 10).

Saving Irrigation Water

Similarly, the efficiency with which irrigation water is used could be greatly improved. A wiser choice of irrigated crops could boost production considerably. According to Sandra Postel, bad irrigation management and inappropriate choice of crops result in much of the world's existing irrigated area being underused, yielding far below its potential and, in some cases, failing to enhance food and income security for those who most need it – the rural poor.[13]

The assumption that the use of water for irrigation will increase linearly with population has also been challenged. As with any resource, growth in use inevitably slows when supply constraints become apparent. The relentless advance of irrigation in the 1960s and 1970s, for example, slowed considerably in the 1980s when expansion averaged only about one per cent a year compared with 2–4 per cent in the previous two decades.

This was partly because the best sites for new reservoirs and dams had already been used, and new projects therefore became relatively more expensive. Between one-quarter and one-half of the total run-off on every continent is now stabilized by dams and reservoirs. Access to new areas where run-off can be controlled is often difficult, or would have such a major impact on existing populations, forcing a large-scale and always unpopular relocation of people, that a number of proposed schemes have been abandoned. The number of new dams inaugurated in the world peaked at about 500 a year in 1970, and has now fallen to below 400 a year. But social and economic factors have been important, too: the development of irrigation schemes has been curtailed by low commodity prices, expensive energy, and a poor economic outlook.

Great economies are possible with irrigation. Experience in the United States has shown that while normal irrigation usage is only about 45 per cent efficient, farmers who use more expensive pumped ground water often achieve efficiencies of more than 60 per cent. Economics is the spur to their achievement – as it could be to many others. Israel has gone one better; its trickle irrigation schemes, in which water is trickled to crops down furrows, and the ground is never flooded, have reached efficiencies of 95 per cent.[14]

The Social Dimension

What matters most is not whether a country has enough water to satisfy its theoretical needs; what matters is how people, particularly those in rural villages and farms, can best adapt their use of water to provide the greatest benefits from what is available.

Even many developed countries are far from doing this. Water supplies in the United States, for example, are copious.

Yet 27 per cent of the country's irrigated land has been damaged by salinization and 21 per cent of its irrigated area is fed from underground aquifers that are being depleted faster than they are being recharged.[15] Even in 1980, the State of California was using 41.3 cubic kilometres of water a year, the equivalent of 112 per cent of its reliable run-off.

As in China and India, the water table is falling, in some areas fast, and those who rely on such supplies are living on borrowed time. Their use of water is unsustainable. Those who rely on irrigation in the arid south-west of the United States are faced with a water crisis that affects their daily lives, and which is likely to worsen. To alleviate the problem cities from Utah and Colorado to Nevada and New Mexico are buying up farmland so that they can acquire the water rights that go with it. Phoenix, in Arizona, for example, bought nearly six million hectares of farmland for $29 million in 1987 just so that it could satisfy its inhabitants' burgeoning demand for water.

At the other extreme, Malta, with a per capita supply of run-off equivalent to only 60 cubic metres per person a year (and no rivers flowing in from neighbouring countries), is further past the "water barrier" than any other country is likely to go. Yet life on Malta goes on. Indeed, life styles, while not luxurious, are far superior to those of many of the inhabitants of even some well-watered developing countries. Malta solves its problems by the careful use of what water it has, by desalination and by importing water – and food – when it needs to.[16] There is no reason why such solutions should not be applied in at least some water-poor areas in developing countries. Water self-sufficiency, like food self-sufficiency, can be a false goal.

Such examples show two things. First, even countries where overall water supplies are copious can suffer from severe local shortages. And, conversely, even countries that are perilously short of water can, sometimes, manage to achieve decent life styles for their inhabitants. Thus the basic issue is not a technical one. It is, as Malin Falkenmark says, ". . . in reality a problem of incentives, social controls, ability to handle complexity as well as social pressures and conflicts".[17]

But this does not mean that that there is no problem. On the contrary, there is every indication that many countries will soon be faced with hard social and economic choices occasioned primarily by water shortage.

The Real Crisis

Thus there is no room for complacency about the world water situation. There is, indeed, a crisis looming, and one that will require radical changes in our approach to development if its effects are not to cause much unnecessary suffering. The following summary, by Sandra Postel, illustrates the situation dramatically.

Country/region	Water situation
N. and E. Africa	Ten countries likely to experience severe water stress by 2000; Egypt, already near its limits, could lose vital supplies from the Nile as upper-basin countries develop the river's headwaters.
China	Fifty cities face acute shortages; water tables beneath Beijing are dropping 1–2 metres per year; farmers in the region could lose 30–40 per cent of their supplies to domestic and industrial uses.
India	Tens of thousands of villages now face shortages; plans to divert water from Brahmaputra River have heightened Bangladesh's fear of shortages; large portions of New Delhi have water only a few hours a day.
Mexico	Groundwater pumping in parts of the valley containing Mexico City exceeds recharge by 40 per cent, causing land to subside; few options exist to import more fresh water.

Middle East	With Israel, Jordan and the West Bank expected to be using all renewable resources by 1995, shortages are imminent; Syria could lose vital supplies when Turkey's massive Ataturk Dam comes on-line in 1992.
Soviet Union	Depletion of river flows has caused the volume of the Aral Sea to drop by two-thirds since 1960; irrigation plans have been scaled back; high unemployment and deteriorating conditions have caused tens of thousands to leave the area.
United States	One-fifth of total irrigated area is watered by excessive pumping of ground water; roughly half of western rivers are over-appropriated; to augment supplies, cities are buying farmers' water rights.

The first signs of water stress usually involve conflicts between different types of water user, notably farmers, industry and urban populations. In the well-ordered United States, the conflict takes the form of cities buying out farmers' rights to the water on their land, and then piping it to the city supply. And many farmers are willing enough, if they make more money selling the water than using it to spray their crops. Prices can be high. In Colorado farmers have sold water at a rate equivalent to providing a US household with water for two years for between $3000 and $6000. Water at that price surely spells the end of the era of cheap or even free water – at least in Colorado.

Elsewhere, the conflict can take the form of bitter battles. In this game, industry is usually the winner. The reason is a matter of simple economics. In 1980 in California it was estimated that every cubic kilometre of water used to water crops produced an added value of $75 million; the same cubic kilometre used by industry produced an added value of $5 billion – more than 65 times as much. Whether by coincidence or not, Chinese planners also calculate that the water used in industry is, in crude economic terms, 60 times more productive than water used in agriculture. It is hardly surprising, then, that industry fights for its water, even to the exclusion of surrounding farmers who usually own the water rights that go with their land.

The Indian pulp and paper industry has expanded four-fold over the past decade or so; it uses water heavily, and guards its "rights" to water fiercely. A pulp factory at Nagda, on the Chambal River, draws a 114,000 cubic metres a day for its operations from two reservoirs which it built specially for the purpose. But, says, Babulal Bharatiya, whose farm is next to the reservoirs, "During the summer, when water is very scarce, the factory does not allow the farmers to irrigate their land. Security guards, who keep a round the clock vigil, even assault the villagers".[18]

If anyone still doubts that a water crisis is pending in the drier parts of the world, they should visit the Nile, a river that delivers some 90 cubic kilometres of water to the Mediterranean a year. Or used to. Since the Aswan High Dam was filled in the mid-1970s, in some years the entire flow of the Nile is used up before it reaches the sea.

Notes

1 Malin Falkenmark, "The Massive Water Scarcity Now Threatening Africa – why isn't it being addressed?" *Ambio*, vol. 18, no. 2, 1989. Falkenmark actually defines her categories in terms of the number of people who rely on a supply of one million cubic metres of water a year. Thus the water barrier occurs when more than 2000 people must live off a supply of one million cubic metres a year. The original figures have been transposed for the sake of consistency with other figures used in this book, which are given in terms of annual per capita consumption.

2 World Resources Institute, *World Resources 1986 – an assessment of the resource base that supports the global economy* (New York: Basic Books, 1986).

3 US Council on Environmental Quality and the Department of the State, Gerald O. Barney (Director), *The Global 2000 Report*

to the President: entering the 21st century (Washington, DC: US Government Printing Office, 1980).

4 Falkenmark, *op. cit.*

5 UN Economic Commission for Europe, *Long-term perspectives for water use and supply in the ECE region.* (New York: United Nations, 1981, Sales No. E.81.II.E.22).

6 See, for example, Gilbert F. White, "Water Resource Adequacy: illusion and reality" in Julian L. Simon and Herman Kahn, *The Resourseful Earth: a response to Global 2000* (New York: Basic Books, 1987).

7 UNEP, *Environmental Data Report* (Oxford: Blackwell and UNEP, 1989).

8 Robert P. Ambroggi, "Water", *Scientific American*, September 1980.

9 Peter Rogers, "The Future of Water", *The Atlantic Monthly*, July 1983.

10 Ambio Round Table Discussion, "Water-Related Limitations to Local Development", *Ambio*, Vol. 16, no. 4, 1987.

11 Sandra Postel, "Increasing Water Efficiency" in Lester R. Brown et al. (eds.), *State of the World 1986* (New York and London: Norton, 1986).

12 World Resources Institute, *World Resources 1986 – an assessment of the resource base that supports the global economy. Op. cit.*

13 Sandra Postel, *Water for Agriculture: facing the limits* (Washington, DC: Worldwide Institute, 1989, Worldwatch Paper 93).

14 Peter Rogers, *op. cit.*

15 Sandra Postel, *op. cit.* The definition of overpumping used to produce these figures was areas where the water table was falling by at least 15 centimetres a year.

16 Marseilles has become the water-exporting harbour of the Mediterranean. See G. Power, "Water Towing, Suppliers Search for Shuttle Markets", *World Water*, February 1984.

17 Malin Falkenmark, personal communication.

18 N. D. Jayal, "Destruction of Water Resources – the most critical ecological crisis of East Asia", *Ambio*, vol. 14, no. 2, 1985.

Chapter 6

Water and Development

Water is the source of all life.

The Koran

The most authoritative non-technical report of the planet's ailing environment was published in 1987. Advertising itself on the back cover as "the most important document of the decade on the future of the world", *Our Common Future* was four years in the making. Compiled by the World Commission on Environment and Development, which was set up by the United Nations in 1983, the work involved a series of public hearings and special meetings held in 15 cities, involving more than 500 written submissions amounting to more than 10,000 pages. The cooperation of some 1000 individuals and institutions is acknowledged in an annex.

The book starts boldly:

> The disasters most directly associated with environment/development mismanagement – droughts and floods – affected the most people and increased most sharply in terms of numbers affected. Some 18.5 million people were affected by drought annually in the 1960s, 24.4 million in the 1970s. There were 5.2 million flood victims yearly in the 1960s, 15.4 million in the 1970s . . . The results are not in for the 1980s. But we have seen 35 million afflicted by drought in Africa alone and tens of millions affected by the better managed and thus less-publicized Indian drought. Floods have poured off the deforested Andes and Himalayas with

increasing force. The 1980s seem destined to sweep this dire trend on into a crisis-filled 1990s.[1]

But this promise is not fulfilled. The book contains no special chapter on water, and includes no direct analysis of water issues. The index provides 21 separate page references to water but all treat the issue *en passant*.

This lamentable neglect of one of the planet's fundamental environmental issues brought quick action from the International Water Resources Association. At the Fourth World Congress on Water Strategies, held in Ottawa from 30 May to 3 June 1988, the IWRA drafted a statement on the WCED's shortcomings:

> The Commission pays no attention to the galloping and multi-dimensional water scarcity now developing in Africa . . . It is remarkable that there are sub-chapters on the oceans, space and the Antarctic but not one on fresh water, the blood and the lymph of the geophysiological system that we call the biosphere . . . The absence of a credible discussion of sustainable development as seen from a water perspective caused great concern, even dismay, at the World Water Congress.[2]

The full text of the IWRA statement is given in appendix B.

Unhappily, the WCED report had many precedents. *North-South: A Programme for Survival* (known as the Brandt report) was published in 1980. Its 304 pages included just one devoted to water, and that treated soil and water as if they were the same resource.[3] "Governments", the report admonishes sternly (and ungrammatically), "should find ways to transfer workers in the slack season to improving the land with such activities as building, fencing, drainage or small irrigation construction". Talk like this appears already to come from another era in development work, even though it was written only 10 years ago. Even so, to bracket fencing with small-scale irrigation construction is indicative of the report's attitude.

There is a long history of what Malin Falkenmark calls "water paralysis" in development thinking and action. It stems, she claims, from a poor understanding of the importance of the water cycle to security and well-being in the tropics on the part of the donors and experts who make so much of the running in the current development climate. They live in water-rich societies. Where their plans and projects embrace areas where water is scarce, they tend to think in terms of the technological fix: how to make available the water needed for a new village, an urban centre, an irrigation project. Rarely do they follow through with any analysis of the broad implications of water security, or of the effects of one project on another.

Unhappily, these basic attitudes are often instilled into the hydrologists and hydraulic engineers from developing countries – many of whom, of course, must train in industrial regions for lack of appropriate courses at home. As Professor Arie Issar, of the Ben Gurion University of the Negev, writes:

> When they return to their own countries, they transplant both the knowledge and the attitudes, seldom having the background knowledge or the courage to modify such methods and doctrines, not to speak of inventing new ones. Magnificent dams standing high and dry above dusty riverbeds, or reservoirs silted up or even breached by unexpected desert floods, are sad but thought-provoking sights in many arid regions.[4]

Describing the sophisticated water management civilization that used to thrive in the arid regions of Tunisia and other north African countries, Slaheddine El Amami, Director of the Research Centre on Rural Engineering in Tunis, is even more emphatic. Why, he asks, has such a workable and efficient process been allowed to slip into decay and ruin? Why could not the best of the old be combined with the best of the new, to provide a synthesis of ideas that would have served the country even better? Instead, the old systems have been swept away and replaced by new, centralized ones that depend on the construction of one or two huge dams. The result has been erosion, lowering of water tables, salinization and rural poverty. He writes:

The modern versus traditional antithesis spread by the colonizers, the onslaught of imported technologies, the modernist language of the national political e/lite – all of these have helped to discredit local know-how and traditional technology, regardless of its suitability to local conditions. Moreover, engineering schools have played a tremendous role in the dissemination of the dominant technological model. They serve as a "mould" to turn out engineers alienated through high technology. They are encouraged to scorn or discard every form of local technological heritage by their instructors who spread such preconceived notions as "the lowest kinds of technology require a vast labour force".[5]

Anyone who thinks modern technology does not also require a vast labour force should watch the construction of a modern dam.

The slippery way in which water issues in one area quickly flow through to impact on other areas is mainly ignored in modern water management schemes. Even northern ecologists tend to think only of the need to get water into the root zone of plants; they regard water as a nutrient, like fertilizer, something that must be supplied as needed. They neglect the effects of water penury in societies accustomed to carrying their daily supplies, a few litres at a time, from distant wells and streams. And this is the crux.

For Want of a Bucket

To those who are professionally involved in development, the failures are often more apparent than the successes. Development writers have become habituated to trying to tease cautious optimism from the successes – and, more often, drawing instructive lessons from the failures.

But there comes a time to pause. Why is it so difficult to make development work? The goodwill is there; adequate money is often available; enthusiastic and trained staff are not hard to find.

One commonly cited reason for the failure of development

efforts is the Westernized, developed world model that surrounds many development efforts. Since development itself is a Western concept, this is not entirely surprising. Nor does it necessarily ensure failure. Yet the attitudes to water of temperate society, as we saw in the last chapter, are very different to those of people in water-poor areas. And while this certainly does not explain development's poor success rate, it equally certainly contributes to it.

In 1983 I visited a small-scale project in the Arusha district of Tanzania. The goal – that of building village grain storage huts, so that food would not rot before the government agency came to purchase it, often more than a year after the harvest itself – was praiseworthy enough. It promised to increase farmers' incomes, to reduce foreign spending on imported food, and to give people more to eat.

It was an arduous day's drive from Dodoma, the country's political capital, to the village concerned. The last score of kilometres was on muddy and sometimes flooded dirt tracks. The last kilometre was by foot. The village lay nestling in wooded country, not far from the river valley. And the first sight I had of the villagers was a mournful group, standing a respectful distance from a concrete slab shaded from the sun's fierce heat by a few planks. On it lay a child dying from malnutrition, or one of the many diseases caused by malnutrition.

Had I had the photojournalist's eye for the telling photograph of African hardship, I could have stepped in. But this was my first real sight of African suffering. I stood back, and my guides hurried me on.

They took me to see the site of the project itself. There stood the village grain store – or, at least, there stood its foundations, marking the culmination, apparently, of two years' work and two years' funds from the Arab donor.

There was nothing more: just a concrete foundation. I was led down a field to where the raw materials for the walls were to have been fabricated. There were eight piles of rotting bricks. Bricks, that is, that were not properly fired – or, in this case, sun-dried – and had then been eroded by wind and rain. A few score could perhaps have been used. Thousands were crumbling. And many other thousands had been reduced to powdered clay. No one could have built a village grain store from them. They were, perhaps, enough to build a latrine shelter that could be used by half a dozen families.

There were no men in the village that day. They had travelled, on foot, a dozen kilometres to the nearest large farm (run by Europeans) to help with the harvest at a daily wage of a dollar or two. This was the lean time – the time between the harvest coming in, and the time after the last harvest supplies had run out. Hence the death on the outskirts of the village. But this was precisely the time for which the village grain store had been planned to cater.

My guides were quick to explain. "There is nothing wrong with the project", they said. "On the contrary, the villagers have been enthusiastic. Otherwise there would have been no bricks at all. But we had no foreign exchange provision in the project to buy the two things that the villagers could not make: cement and buckets. And without buckets the villagers cannot carry the water they need even to make their own form of mortar. So the bricks have been left where they were made, and have rotted. Six months ago we asked for a small additional grant, in dollars, to buy what we need. We have the dollars. But we cannot buy buckets in Tanzania. They have to be imported. And government restrictions on imports are now so tight that we are still awaiting clearance to buy the buckets."

But for the want of a bucket that African child might have lived (he had died by the time I left the village). But for the want of a score of buckets that Tanzanian village would have had its grain store. And once erected vertically, the sun-dried bricks would have withstood the ravages of wind and rain for perhaps a decade or two. As it was, a few thousand dollars had been wasted. And a few African lives.

Water Blindness

The moral of this story has, of course, to do with water. Not that water itself was the key issue. The village in question lay but a kilometre or so from a good supply, that ran most of the year. Water scarcity was not the issue. Water blindness was. No one had thought in the Arab development agency that financed the project, at the United Nations agency that was executing it, or even among the Danish volunteers who were directing it, to allow for the purchase of buckets. And buckets were the breaking point.

If this story seems trivial, it is not. Many thousands of development projects have foundered for similar reasons. Even large-scale water projects, that aim to furnish thousands of farmers with irrigation canals, neglect the obvious. Development experts have become blind to the issue of water. Their northern upbringing, which comes from lands where water is copious, where taps are everywhere, and where the cost of water is less, much less, than peanuts, leads them to omit consideration of the one really crucial factor: water.

It is time this farce was brought to a close. The previous chapters have demonstrated the critical role that water has to play in the lives of the rural poor in developing countries. It is time to make specific provision in future projects for water needs. And that means not only ensuring that those who will ultimately benefit have the water they need to fulfil the project, but also that they have the tools they need to get the water from where it is to where it needs to be.

Every project needs water. Few mention the fact. Every project costs money. All projects cost water – but they rarely say so. They measure performance in terms of population density per hectare of arable land, of increased yield per hectare and of numbers of people served by new medical centres, markets, fishing fleets and conservation projects. They never compute the cost in terms of cubic metres of water per building erected or the numbers of villagers who may benefit.[6] They should. And if the emergencies of the 1980s are to become a thing of the past, they will have to. Otherwise, calls for emergency food supplies and for disaster relief will go on mounting just as they have over the past decade. It is time to take water seriously. And this means making specific provision for water in *every* development project in the future, and for computing the cost of every project in terms of water saved and water gained.

The problem has nothing to do with the number of projects that are devoted at least partially to increasing water availability, and particularly to providing water for irrigation – though it has much to do with their scale and their appropriateness. As Chapter 8 shows, the industrialized regions have developed their own particular solutions to water management issues. They mostly involve technical fixes and gargantuan developments that encompass whole rivers, regions, countries and even sub-continents. They are expensive, involve difficult technologies and are conceived in regions where rainfall is high and barely seasonal, and where evaporation rates are low.

Exporting such schemes to tropical countries where evaporation rates are high, and rainfall is either scant or excessive for three months and then negligible for the rest of the year, has proved more than difficult; in many cases, it has proved disastrous.

In fact, few developed countries have anything to offer developing countries in terms of improved water management. The one area that ought to provide insights – south-west United States – is conspicuously short of potential. Demand in that area is simply too high, and attempts to increase supplies through groundwater pumping, desalination and the long-range transport of water have been a great deal less than successful. The only remaining solution to the problem in the south-west of the United States is to reduce water use.

Developing countries will do best, argues Malin Falkenmark, not to ask the familiar questions of the West: how much water do we need and where can we find it? Instead, they should ask: how much water is there, and how can we best benefit from it? Few do.

A recent assessment of water resources in the Arab

League countries by the Islamic Network on Water Resources Development and Management (INWARDAM) illustrates the point. The assessment assumed a basic demand for water of 55 m³ per head per year for domestic water use, plus 1150 m³ needed to provide an average daily diet of 3000–3500 kilocalories (375 kg/year of fruit and vegetables, 35 kg/year of meat and 125 kg/year of cereals). The total, of 1205 m³ of water per head per year, was called the lower limit of water requirements.

In 1985 nine of the Arab League countries could meet this basic level of water demand; twelve could not. But, overall, the average per capita supply was well in excess of the minimum, at 1750 m³. By the year 2000, as a result of population growth, overall availability was estimated to fall to just under the minimum, to 1108 m³. And by the year 2025, supply would lag far behind demand, with only 536 m³ available. By then the region's deficit would amount to more than 421 billion m³ of water a year – more than the region's total supply. Only Iraq, Mauritania and Lebanon would be able to meet even the lower limit of water requirements.

With considerable lack of conviction, the assessment goes on to suggest some solutions, including "wise and flexible water policies and strategies", and the reuse of sewage water and brackish water, and salt water desalination.

However wise, however flexible, neither new policies and strategies, nor the technologies mentioned above, are likely to solve the problem. One of the key issue in the politics of water resources in the semi-arid zones is the disparity between the wet and the dry seasons. Somehow, more water must be harvested from the wet season to lubricate life in the dry season. Inevitably, this means more dams and reservoirs. Means will have to be found of building smaller structures that have less devastating effects on local populations.

Further in the future, long-term water storage may be increasingly underground. Israel is already using underground aquifers to store its winter rains, and other semi-arid countries will soon follow suit.

Finally, many semi-arid countries will have to ask hard

questions about food self-sufficiency. The Arab League statistics speak for themselves: 1150 m³ a head a year to produce food, 55 m³ a head a year for domestic water use. Growing food uses a great deal of water. Many semi-arid countries could transform their economies if they imported more food, grew less food and exported other, higher value and preferably less water-intensive products. Lateral solutions of this kind will be the key to the water security of many of the world's 50 or so poorest and driest countries.

Notes

1 The World Commission on Environment and Development, *Our Common Future* (Oxford and New York: Oxford University Press, 1987).

2 International Water Resources Association, *Sustainable Development and Water: statement on the WCED Report "Our Common Future"* (Ottawa: 1988).

3 Independent Commission on International Development Issues, *North-South: A Programme for survival* (Cambridge, Massachussetts: MIT Press, 1980).

4 Arie Issar, "The Abuse and Use of the Hydrological Cycle", *Impact of Science on Society*, no. 1, 1983, Managing Our Fresh-Water Resources.

5 Slaheddine El Amami, "Changing Concepts of Water Management in Tunisia", *Impact of Science on Society* no. 1, 1983, *Managing Our Fresh-Water Resources*

6 Arie Issar, *op. cit.*, puts it thus: "A reader might take any feasibility report on an agricultural development project in an arid region. It is highly probable that he will find in the report that the economic calculations are based on the increase of production per unit area of land – be it acre, hectare, dunam or fedan. Seldom will the reader find a study where income is calculated per unit volume of water."

7 *Newsletter of the Islamic Network on Water Resources Development and Management* (INWARDAM). July-September 1990, Issue No 10.

Chapter 7

Water and International Conflict

"Water – not oil – will be the dominant resource of the Middle East . . ."

M. Falkenmark
Ambio, *vol. 18, no. 6, 1989*

Water is a unique natural resource not only because it cannot be created or replaced but also because it moves. Unlike coal, oil, forests and soil, water flows, from mountain top to plain to the sea and, inevitably, from one country to the next.

The way rivers are used in one country can therefore have far-reaching effects on all other downstream countries. Large-scale irrigation or the damming of a major river can bring prosperity to a country where water availability is unreliable; but the same scheme can threaten countries further down the river with ecological and economic disaster. Upstream countries can unilaterally affect the amount of water reaching countries further downstream – either reducing it, and causing domestic or agricultural water shortages, or increasing it, and causing flooding. Upstream countries that pollute their rivers – and that effectively means *all* upstream countries – can also reduce the amount of water which the downstream country can use for drinking or for agriculture.

Conflicts over access to, or the quality of, fresh water began early in history: the Mesopotamian cities of Lagash and Umma were in dispute over water as early as 4500 BC. Today, the sharing of water resources contributes to many international disputes, particularly in countries where water is scarce, and its use unregulated by treaty.

Competition for water is fierce in many arid areas, and getting fiercer. Elsewhere, densely packed river valley communities are being put at risk from floods produced by deforestation high up in the mountains whence their water supplies originate. Deaths from these floods can number in the tens of thousands.

The political importance of water as a resource has few boundaries. And countries that need it pay scant attention to political boundaries. Water, of course, is the key to economic activity throughout the Middle East, and assured supplies of fresh water have already been used as bargaining tools in attempts to achieve a Middle East peace settlement.

The Roots of Water Conflict

The potential for conflict is enormous. Globally, 47 per cent of all land falls within international river basins, and nearly 50 countries on four continents have more than three-quarters of their total land in international river basins. Two hundred and fourteen basins are multinational, including 57 in Africa and 48 in Europe.[1]

In human terms, this means that almost 40 per cent of the world's population lives in international river basins. These two billion people are dependent on the cooperation of all the countries sharing the basin for a guaranteed water supply of consistent quality, and for their environmental stability.

Thirteen river basins are shared by five or more countries. Ten are shared mostly by developing countries, and have few or no treaties to regulate water use; and a number are in areas where water is otherwise scarce. It is not surprising that many have a history of international tension, particularly the Jordan and Euphrates in the Near and Middle East; the Ganges in Asia; the Nile in Africa; and the Colorado and Rio Grande in North America.

In developed countries, many international agreements have been drawn up to regulate shared basin areas and, as a result, use of shared water supplies is more rarely a source of international dispute. Europe, for instance, has four river basins shared by four or more countries, but these are regulated by no less than 175 treaties.

RIVERS AND LAKES WITH FIVE OR MORE NATIONS
FORMING PART OF THE BASIN

Danube	12	Mekong	6
Niger	10	Lake Chad	6
Nile	9	Volta	6
Zaire	9	Ganges-Brahmaputra	5
Rhine	8	Elbe	5
Zambezi	8	La Plata	5
Amazon	7		

Source: P.H. Gleick, "Climate Change and International Politics: Problems Facing Developing Countries", in *Ambio*, vol. 18, no. 6, 1989.

An absence of regulation on shared water resources is common in developing countries. Africa, in comparison to Europe, has a vast and complex system of river basins: 12 are shared by four or more nations, but only 34 treaties regulate their use. In Asia, only 31 treaties have been drawn up to regulate the five basins shared by four or more countries.

A lack of international agreements over water increases the potential for dispute. Competition for water is also intensifying as limited resources come under pressure from increasing populations, particularly in developing countries, and climate changes increase water scarcity in arid and semi-arid regions. Many water-scarce regions are located in the shared basin of a major river system; the fact that these regions are often areas of political instability increases the potential for international dispute still further.

Speaking of growing pressures on fresh water supplies in 1984, Dr Mostafa Tolba, Executive Director of the United Nations Environment Programme, went straight to the point: "National and global security are at stake", he claimed. "Shortages of fresh water worsen economic and political

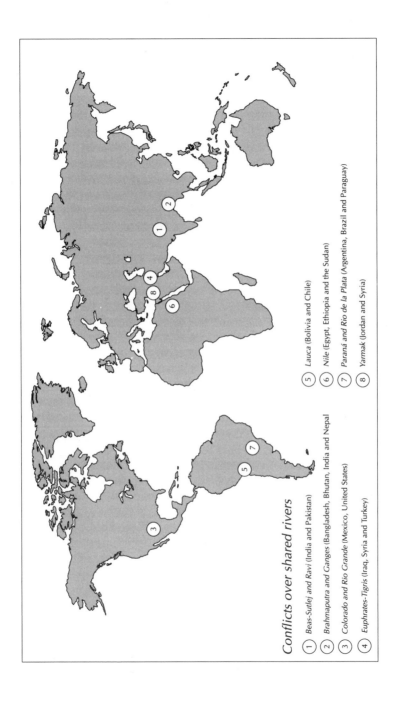

Conflicts over shared rivers

1. *Beas-Sutlej and Ravi* (India and Pakistan)
2. *Brahmaputra and Ganges* (Bangladesh, Bhutan, India and Nepal)
3. *Colorado and Rio Grande* (Mexico, United States)
4. *Euphrates-Tigris* (Iraq, Syria and Turkey)
5. *Lauca* (Bolivia and Chile)
6. *Nile* (Egypt, Ethiopia and the Sudan)
7. *Paraná and Rio de la Plata* (Argentina, Brazil and Paraguay)
8. *Yarmak* (Jordan and Syria)

differences among countries and contribute to increasingly unstable perceptions of national security."[2]

All countries that rely on water originating outside their territory are dependent, in the absence of treaties, on the goodwill of upstream countries. Goodwill cannot be guaranteed, especially under conditions of scarcity. Downstream countries also often find themselves in a weak position to negotiate formal arrangements. This insecure situation is intensifying as water scarcity increases, and control over the national water supply often becomes a political issue, particularly in countries already impoverished by the over-exploitation of natural resources.

In contrast, countries that control the water resources of neighbouring nations can wield formidable power. As water resources have become increasingly scarce, countries have not hesitated to take action to ensure their own supply, even at the expense of downstream countries. Such actions make international disputes inevitable and increase the likelihood of armed conflict.

Conflict in South Asia

Since independence in 1947, the conflicting water needs of India and Pakistan have made control of the shared resources of the Beas-Sutlej and Ravi rivers a highly-charged political issue.

These rivers rise in the Himalayas, and flow through the predominantly Sikh state of Punjab in India, into Moslem Pakistan. To the south of Punjab are the predominantly Hindu Indian states of Rajasthan and Haryana which do not have their own river system. In 1948 India diverted the course of the Beas-Sutlej and Ravi rivers so that they no longer flowed into Pakistan. This happened in the spring growing season and threatened Pakistan's agricultural output for the whole year. The dispute was settled only through the intervention of the World Bank.

Punjab water was an important contributing factor in conflicts such as the 1965 Indo-Pakistan war, and the Indian storming of the Sikh Golden Temple in Amritsar in 1984. The Punjab presently gets about 40 per cent of India's agreed share of the Beas-Sutlej-Ravi River water, and the Sikhs' main political group is asking for a greater amount of the river water to be made available to the Punjab for irrigation. The water issue has become inextricably linked with the wider religious and political demands that led to the violent confrontation at the temple in Amritsar.

Potential river conflict areas in South Asia

The basin of the Ganges and Brahmaputra rivers is shared by four countries: India, Bangladesh, Nepal and Bhutan, and

250 million people depend on the coordinated management of the basin for their survival. However, poor management has resulted in severe problems for downstream countries. There has been widespread deforestation of the highlands of Nepal and Bhutan where the river rises. This in turn has caused soil erosion and increased deposition in the river. Every year 240 million cubic metres of Nepalese topsoil are carried downstream to India and Bangladesh, where they are threatening the viability of dams and irrigation projects. Deforestation has also led to erosion and faster run-off which have increased the areas prone to flooding in the two countries. Four years of particularly disastrous flooding in the 1970s caused the death of some three-quarters of a million people in India and Bangladesh, and damaged billions of dollars worth of property.

Lack of coordinated management has caused problems since large-scale water management projects became feasible for the individual countries in the basin area. A dispute between India and Pakistan flared up in 1948 over India's decision to build a barrage across the Ganges at Farakka, 19 kilometres from the East Pakistan (now Bangladesh) border. India wanted to divert the waters into her own river system to avert the silting up of Calcutta, the country's major port, and to maintain water supplies for 100 million people resident in the basin area. Pakistan, however, argued that although this project was to be carried out beyond its national boundaries, it was an infringement of the country's sovereignty because it would prevent Pakistan carrying out projects within its own territory. These included an irrigation scheme covering almost 2 million acres, which would have increased food production and significantly reduced damage caused by flooding and drought in the area. The conflict of interests was not finally resolved until nearly 30 years later, in 1977.

Conflict in South America

Thirty-six of South America's rivers flow through more than

one country, and the region's shared river basins are inhabited by about 100 million people. Most rivers flow from the Andes down to flat plains which extend for several thousand kilometres. These areas are particularly prone to flooding, and any increase in the volume of river water can inundate huge areas of prime farmland. This makes countries extremely sensitive about activities in other countries further upstream.

Itaipu Dam on the River Paraná

The basin of the Rio de la Plata system, which includes the Paraná River, is shared by Argentina, Bolivia, Brazil, Paraguay and Uruguay. Brazil is in a dominant position as the upstream country, and the effects of its activities are multiplied in downstream countries because the river basin covers a much larger proportion of these countries than it does of Brazil. This has led to conflict.

Brazil has carried out a number of water projects, such as building dams and hydroelectric works, without consulting the other affected countries, and has sometimes deliberately withheld information on downstream environmental impacts.

Paraguay and Brazil collaborated on the construction of the Itaipu Dam on the Paraná River, which is part of the Rio de la Plata system. The dam has created a lake 200 kilometres long, which holds enough water to flood almost all of north-east Argentina. Although Argentina is the country most affected by the plans, it was not part of the joint venture. Relations between Brazil and Argentina soured over the affair, and opponents of the dam, including Brazilians, claimed that the development bore less relation to the country's energy needs (the dam can generate 20–30 per cent more energy than Brazil currently uses) than to Brazil's "militant posturing" towards neighbouring countries.[3]

On the other side of the Andes, Chile and Bolivia have been in dispute over the waters of the River Lauca since 1962, when Chile – where the river has its source – diverted some of the river water for a major hydroelectric and irrigation project. Bolivia, the downstream country, warned Chile that this would be seen as an act of aggression. Bolivia claims that the salinity of Lake Coipasa, into which the Lauca flows, has increased as a result of the diversion; that humidity in the basin area has decreased due to the reduced river level; and that the reduced volume of water in the river has caused water shortages in a number of cattle farming areas.

Two days after the diversion, Bolivia cut off diplomatic relations with Chile, and in September 1962 took its case, unsuccessfully, to the Organization of American States. In 1968, new attempts were made to reach agreement, but these also failed and, since 1979, relations between the two countries have deteriorated further.

Conflict Between Mexico and the United States

The river basins of the Colorado and Rio Grande are shared by the United States and Mexico, though only a limited framework of treaties and agreements on water use has been developed. This is mainly because water is a national resource in Mexico and its use can therefore be regulated by the government; whereas in Texas water belongs to individual landowners who have the right to deplete it without limit. Landowners are unwilling to participate in any water strategy programme that curtails their absolute control of water on their land. The water resources of the border area have been a source of conflict between the two countries since the turn of the century.

In both Mexico and the United States water is taken from underground aquifers to be used in irrigation. The area of irrigated land has increased by some 60 per cent over the past 30 years,[4] and this has caused a severe lowering of the water table which is likely to continue for the foreseeable future. Farmers on both sides of the border are short of water as a result.

Mexico blames the falling water level on Texas for increasing the number of wells on its side of the border, and has also accused the United States of reducing the quality of river water entering Mexico from the United States. And certainly a number of US water-development projects have failed to take into account the serious downstream effects in northern Mexico.

In the early 1960s, crops in the Mexicali Valley in northern Mexico began to show signs of salt poisoning. This region has been made highly productive through irrigation, and the water for the irrigation system comes from the Colorado River which originates in the United States. The salinity of the irrigation water nearly doubled in 20 years, affecting some 80 per cent of the Colorado delta's arable land in Mexico, and making some 14 per cent of the land too saline for cultivation. Calculations showed that agriculture might be wiped out altogether in the valley by the year 2000.

The causes of the increase in salinity in the river were complex but all related to increased irrigation. A large dam at Glen Canyon in the United States, used to hold back water for irrigation, was effectively reducing the fresh water flow downstream to Mexico as a result of the withdrawals made from it. Suplus irrigation water was dissolving salts as it moved through underground layers of rock on its way to the Colorado. And an irrigation project in Arizona was draining highly salty water directly into the river.

Eventually the United States agreed to build a canal to carry some of the saline water from the irrigation scheme to the Gulf of California, and to build a desalination plant in Arizona. Partly due to these measures, the salinity level of the Colorado in Mexico has fallen to ten times below its previous level, thus ensuring the continuation of agriculture in the Mexicali Valley.

The Rio Grande, which has its source in Colorado and flows eventually into Mexico, is also a source of conflict between the two countries. The United States refers to an agreement from 1906 as the basis for its unilateral reductions in water supply to Mexico during periods of serious drought or low flow in the rivers in the United States.

In 1978, the US Secretary of State stated that there would be a "diminished and delayed irrigation supply to Mexico" for that year, and that "strictly ex-gratia supplementary amounts might be made available by the US Commissioner".[5] Incidents such as these have increased friction between the two countries over control of the water supply.

The Middle East

Existing tensions in the Middle East add an extra dimension to the difficult problem of sharing limited water resources. Populations are increasing, and water is becoming scarce throughout the region. A 1988 study by the Center for Strategic and International Studies in Washington, DC, concluded that the situation is likely to become so acute that, in the near future, water – not oil – will be the dominant resource of the Middle East.[6] Water scarcity severely curtails agricultural and economic development, and increasing shortages will bring desperate competition for water, heighten existing tensions between countries and increase the potential for armed conflict in the region.

In 1965, a dispute arose about Israel's use of water originating in Arab states. Israel wanted to divert water for its own use, and the failure to reach an agreement with neighbouring countries on the details of the plan led to Israel carrying out the diversion unilaterally. In response, the Arab states in which the water originated planned to divert their rivers into other Arab states, thus depriving Israel of some of its water supply. Regarding this as a serious threat to its security, Israel launched a preemptive strike on Syrian construction sites with military aircraft.

Israel's water-supply problems are exacerbated through over-pumping of the Sea of Galilee which has created salinity problems in the nation's main water source. To avoid the potential water crisis, Israel has developed plans to build a canal and pump water from the Mediterranean to the Dead Sea, and to construct reservoirs above the Jordan valley. This, however, has led to conflict with Jordan, which fears that pumping water into the Dead Sea will waterlog areas of irrigated agriculture in the East Ghor Canal region.

Israel has recently made a water-transfer agreement with Egypt, which originally, under President Sadat, offered Israel 400 million cubic metres per year of fresh water in exchange for a Palestinian solution.[7] Up to a quarter of Israel's water resources is available in the aquifer shared with the West Bank, and any action in this region is likely to increase the divisions and tensions that already exist. Increased water pumping by Israel in the Jordan Valley and along the occupied West Bank is said by Palestinians and Jordanians to have lowered the water table throughout the valley. This deprives Arab farmers of water in the Jordanian East Bank and in the occupied West Bank and, in effect, constrains the economic development of the region and limits the number of Palestinians who can settle there.[8]

Even without the problems arising from long-standing Arab-Israeli differences, many other countries of the Middle East are in dispute over water. Many upstream countries have plans to use the waters of shared rivers and aquifers to fill reservoirs and irrigation systems in order to increase agricultural production. Syria is planning to divert water from the Yarmuk River, Turkey from the upper Euphrates and Tigris, Libya from its shared aquifer and Ethiopia from the Blue Nile.

Near and Middle East rivers

Jordan's water supply is under threat from Syria, which plans to divert 40 per cent of the flow of the Yarmuk River into its irrigation system. This would seriously reduce Jordan's water supply – also used largely for irrigation – and lead to an increase in the salinity of water in the Lower Yarmuk and Lower Jordan. Jordan has therefore signed an agreement with Iraq to transfer water from the Euphrates over the mountains into Jordan.

In turn, both Syria and Iraq are outraged by the action of upstream Turkey, with whom they share the Euphrates-Tigris basin. Turkey reduced the flow of this major river system in January 1990 to allow a huge reservoir behind the newly constructed Ataturk Dam to be filled.[9] The Euphrates is the biggest single source of water for Syria and Iraq, and their supplies are likely to be interrupted for five to eight years.

Although all countries agreed to the construction, Turkey has not set final levels of guaranteed supply to Syria and Iraq. It is unclear whether the 500 m³/second of water which Turkey has agreed to provide to Syria is to be an average, maximum or minimum figure. If a maximum, Syria could see its share of the Euphrates cut by one-third or more.

The combined effect of dams planned for construction in Turkey and Syria could be to reduce the flow of the Euphrates to one-tenth of its usual level by the time it reaches the people living in the Gulf.

Underneath Libya, Egypt, the Sudan and Chad is a massive aquifer and, despite international recommendations that it should be jointly managed by the sharing countries, no agreements have been reached. In 1984, Libya began a massive programme of pumping water from the aquifer for use in Libya. In a speech to mark the opening of this project, Colonel Gadafy reiterated his opposition to the governments of Egypt and the Sudan, and both these countries have objected to the project on the grounds that the extracted water will benefit only Libya, and the pumping will rapidly deplete the shared aquifer and lower the underground water tables in Egypt and the Sudan.

Competition for the waters of the Nile is severe, as all countries sharing the basin are experiencing water scarcity. The Nile basin covers one-tenth of the African continent, and forms part of nine African countries including downstream Egypt and the Sudan, and upstream countries such as Tanzania, Kenya and Ethiopia.

Egypt's increasing water supply problems are due to the country's almost total dependence on the Nile for water. As water needs increased in line with its growing population, Egypt negotiated an arrangement under which the Sudan agreed to provide Egypt with its excess water from the Nile. Because the population of the Sudan is now also increasing, the country is developing large-scale irrigation projects to match the demand for food, and has itself reached the limit of its available water supply. Both countries are therefore turning to upstream countries such as Ethiopia to answer their water needs.

However, Ethiopia is not bound by any agreement with Egypt over water supply, and Ethiopia stated at the 1977 UN Water Conference in Argentina that it was "the sovereign right of any riparian state, in the absence of an international agreement, to proceed unilaterally with the development of water resources within its territory".[10] Ethiopia is presently contemplating diverting as much as 4 billion cubic metres of the Blue Nile into its own irrigation projects, despite opposition from Egypt and the Sudan whose water supply would be severely reduced as a result.

A possible solution to Egypt's problems would be to use Lake Tana in Ethiopia and lakes on the Victoria Nile in Uganda to store water for Egyptian use, as its own Lake Nasser has become insufficient for the country's needs. The project pivots on the governments of Ethiopia and Uganda agreeing to raising the level of their lakes, and flooding habitable areas along the shores. As neither Uganda nor Ethiopia has signed treaties with Egypt, their agreement to such a project cannot be guaranteed.

Egypt's water resources are now under severe stress and its water needs will increase for the foreseeable future. Failure of co-riparian states to appreciate the gravity of the situation and cooperate in water-sharing schemes could tempt Egypt to take a more active interest in the internal affairs of neighbouring countries and threaten the stability of the entire region.

Resolving Potential Conflict

As population levels increase, water-based conflicts are likely to intensify in the future unless international agreements can be reached on the management of shared water resources. Although treaties and management committees effectively avoid or resolve disputes over a number of water courses – those shared by Canada and the United States, for instance – such arrangements have only a localized effect, and cannot reduce the risk of conflict elsewhere.

Wider-ranging regulations on water quality have been introduced by the EC and these apply to all member nations, rather than to a specific disputed area. EC members have cooperated in implementing water regulations on pollution which will safeguard water resources in the future, including those shared by a number of nations. The fact that individual countries are willing to accept transnational legislation which restricts their rights over the use of natural resources within their own territory suggests that an international approach to water management might work in other areas of the world.

Another example of constructive action for management of shared water resources is the plan put forward by Turkey for a "peace pipeline" to transfer fresh water now emptying into the Mediterranean to countries in the Middle East. Two huge parallel pipelines, three to four metres in diameter, are foreseen. The first phase would involve the construction of a western pipeline providing water to Jordan and western Saudi Arabia; a later second phase would involve an eastern line to Syria, Saudi Arabia and the Gulf states. Between six and nine million people could be served by each pipeline, at a level of 400 litres of water per person per day – the present supply level to urban areas in Sweden – if the countries in the area agree to the plan.

Cooperation is also possible over the management of the waters of the Nile. A proposal has been made to set up a

Nile Basin Commission which will allow all nine states in the basin area to participate in the management of shared water resources.

At the global level, there is no established framework of water management. Global agencies do, however, play an important role in training, and the transfer of information and technology for use in water management programmes. The UN is involved in many water-related projects covering, for example, irrigation and the improvement of water quality and supply. The UN does not directly manage these projects, and UN policy states that:

> each nation has the sovereign right to formulate its own environmental policies, provided in the exercise of such right and in the implementation of such policies due account must be taken of the need to avoid producing harmful effects on other countries.[11]

In 1957, the Mekong Committee was set up under UN auspices by Thailand and (former) Laos, Cambodia and South Vietnam in order to harness the waters of the Mekong River – the eighth largest in the world. Since that date, war and political division have thwarted plans for the construction of dams on the main stream of the river in present-day Cambodia and Vietnam, though large-scale dam construction has gone ahead in the Lao PDR and Thailand.

About 50 million farmers depend on the river for their survival, although the river also causes widespread destruction during its seasonal floods. The common aim of all countries sharing the Mekong is to regulate the flow of the river and to use the water in hydroelectric and agricultural schemes. Such plans may gain new impetus from recent political changes in Cambodia.

The Interim Mekong Committee, with support from the UN, is exploring large dam projects in order to produce hydroelectricity and create reservoirs which will increase the amount of fresh water available for agriculture in the dry season. The committee is also investigating schemes for flood control, watershed management and fisheries.[12]

In 1986, the United Nations Environment Programme launched its programme for Environmentally Sound Management of Inland Water Resources (EMINWA). EMINWA's objectives are to provide guidelines to planners on the environmentally sound management and development of inland water systems, and to increase public awareness of the importance of such development.

The top priority is to help countries sharing a river basin to develop their water resources in a sustainable manner and without conflict. This often involves new legal and institutional arrangements – typically a Convention on the management of the shared resource – and the setting up of a river or lake management authority. This authority would oversee the introduction of water treatment technology, the more efficient use of existing water resources, and the implementation of development programmes which do not interfere with the water supply for downstream countries. All these initiatives increase the amount and raise the quality of water in shared basin areas, and lessen both the competition for water and the potential for dispute.

A pilot project is currently being carried out by EMINWA on the Zambezi River, which is shared by eight African countries, to develop an integrated management plan for the whole basin, which can then be adapted for the management of other shared river basins. It involves wide-ranging research into existing conditions in the basin area and the development of a monitoring system to record changes in surface and ground water, sediment, pollution and water quality. On the basis of this information, programmes will be carried out to tackle environmental problems, and improve drinking water and sanitation in the whole basin area.

Although there are currently no major conflicts over the use of the Zambezi basin, demands on the river are bound to increase in the future. As one of the UN experts involved in the basin project puts it, "Any international river needs an international river authority, and for the Zambezi River the need is particularly great. If the countries continue to develop and use the river based only on national priorities, it is bound to come to hard conflict in the not too distant future."[13]

The project on the Zambezi deals with the issues which are at the root of conflict over water. It establishes a framework within which countries can work out the most profitable way of sharing water resources, and sets up a body, recognized by all the countries, which has legal authority to regulate water use in the basin area. This eliminates unilateral action by any of the sharing countries, and thus rules out the short-term threat of conflict. Over the longer term, the project tackles the problem of water scarcity, which is another root cause of conflict, by training people in water management techniques, improving the efficiency of water use, and introducing sustainable deve\pment of water resources.

This is a good – if long-winded – way of \voiding water conflict. If plans like these could be imple\ented in all the major river basins, the potential for confl\t would be greatly reduced. Meanwhile, treaties and agreeme\ts between countries that share water resources are essential\ \conflict is to be avoided in the future. As water becomes m\ \e scarce, they will become more so.

Notes

1   United Nations, *Register of International Rivers* (Oxford: Perg\ \on Press, 1978).

2   Timberlake L. and J. Tinker, *Environment and Conflict* (Lond \: Earthscan, 1984).

3   Timberlake L., and J. Tinker. *op. cit.*

4   McDonald, Adrian T., and David Kay, *Water Resources: issues and strategies* (London: Longman Scientific and Technical, 1988).

5   McDonald A. T., and D. Kay. *op. cit.*

6   Starr R. J., and D. C. Stoll (eds.), *The Politics of Scarcity: Water in the Middle East* (Boulder, Colorado: Westview Press, 1988).

7   Falkenmark M., "Middle East Hydropolitics: Water scarcity and conflicts in the Middle East", *Ambio*, vol. 18 no. 6, 1989.

8 Falkenmark M., *op. cit.*

9 "River of the dammed . . ." *The Observer*, Sunday 25 February 1990.

10 Timberlake L. and J. Tinker, *op. cit.*

11 d'Arge R. C., and A. V. Kneese. "State liability for international environmental degredation", *Natural Resources Journal*, 26, 1980.

12 Birsel R., "Mastering the Mekong", *Development Forum*, Jan-Feb 1990.

13 UNEP, *Safeguarding the world's water* (Nairobi: UNEP, undated, *UNEP Environment Brief No 6*).

Chapter 8

High Tech Solutions

Nothing has happened, however, to alter our view . . . that all funds for large scale water developments should be cut off forthwith.

E. Goldsmith and N. Hildyard
The Social and Environmental Effects of Large Dams

In their efforts to obtain reliable water supplies the developed countries have, literally, moved mountains. They have dammed and flooded, piped, pumped, diverted and canalled, and altered the courses of major rivers. They have removed the salt from sea water, investigated the economics of using supertankers to deliver water like oil, and even studied the prospect of towing icebergs from the Antarctic to melt down for tap water.

Success can be measured on many scales. The environment apart, most of these schemes can be counted successful. They have, by and large, provided the water and hydropower they sought, and they have stimulated industrial and economic growth. If, in places, they have altered ecologies and transformed the landscape for the worse, then – many would argue – the price of progress must be paid somehow.

Flushed with success, the engineers and planners of these schemes have gone forth and preached the word to the unconverted of the Third World. By and large, though still with some exceptions, their missions have been much less successful – certainly in the eyes of the millions of people who have been uprooted from their traditional homes, and provided with new types of job and farming systems for which they were ill prepared. Many new reservoirs and irrigation schemes have spread disease deeper into already sick areas, and signally failed to produce the economic take-off for which they were designed. Nearly all have added to the debt burden of developing countries on which the Third World is now totally unable to pay the interest due.

Even where the effects have been beneficial, many of the gains have been shorter lived than anticipated. High sedimentation rates and falling water tables are testimony to the unsustainable forms of development which have been sponsored.

The high water technology of the West seems poorly suited to both the peoples and the lands of developing countries where, as we have seen, the issues surrounding water are more complex and different in kind to those of the temperate regions.

Diverting the Rivers

The most grandiose engineering schemes known to man concern North America and the Soviet Union. Both have made ambitious plans to divert major rivers that flow north from their territory into the unpopulated Arctic regions. They would have these rivers instead flow south, into the arid continental interiors where the land is dry and the people poor.

In the United States water is already tightly controlled. The waters of the Colorado, for example, are led from the Parker Dam via the Colorado River Aqueduct some 900 kilometres across deserts and through mountains to irrigate California's dusty interior. In Germany, where rainfall is copious, aqueducts of more than 200 kilometres are used to channel water from water-rich areas to water-poor ones. But these schemes would be as nothing compared to the master plan for North America.

The North American plan involves transferring between 136 and 308 cubic kilometres of water a year from seven principal rivers in Canada and three in Alaska. Up to 17 per cent of their run-off would be transferred to central Canada, the south-west United States and northern Mexico. The scheme is known as

NAWAPA – the North American Water and Power Alliance – and was first proposed by a firm of construction engineers in 1964.

The scale of the proposal defies the imagination; there would be 240 new reservoirs, 112 irrigation schemes and 17 new navigable canals or rivers. The largest reservoir would hold 3500 cubic kilometres.

It is probably significant that the United States would receive 61 per cent of the diverted water and Canada and Mexico would receive only 20 and 19 per cent respectively. Canada, which would in effect have to undergo major hydraulic surgery for the scheme, is unlikely to be pleased to see most of the product draining away into the Californian desert.

Furthermore, much of the work and some of the reservoirs would be sited perilously close to fault lines on the earth's crust; seismic disaster would be a serious threat. It is well known – at least to geologists – that making large new reservoirs can trigger earthquakes. According to Omar Sattaur,

> The Koyna Dam in southern India is 103 metres high and its reservoir can hold nearly 3000 million cubic metres of water. By 1963, when the reservoir was less than half filled, the frequency of seismic shocks greatly increased. The epicentres of the shocks were all near the dam or under the reservoir. In 1967, one big seismic shock killed 177 people and injured 2300 others in the village of Koynanagar.[1]

It seems unlikely that the North American scheme will now be carried out. The environmental costs look too high, safety considerations would need much more analysis, and the prospect of getting Canada's agreement to a scheme that gives 60 per cent of the water to the United States yet places the lion's share of the environmental issues on Canada's doorstep are slim. And yet . . . By the next century the United States' south west will certainly need to get more water from somewhere. Problems in Mexico are no less urgent; further expansion of the capital, Mexico City, is now impossible without additional water supplies.

The Soviet schemes are somewhat more realistic; even so,

perhaps the Soviet decision in 1986 to cancel its plan to reverse the flow of three north-flowing rivers, and send an annual 120 cubic kilometres of water 2200 kilometres overland to provide increased irrigation in Central Asia, is even more realistic.

Rivers and lakes in the Soviet Union

The Soviet Siberian scheme was 15 years in the making, and became known as the Soviet Union's "project of the century". It was to have diverted 15 per cent of the annual flow of three rivers – the Ob, the Irtysh and the Yenisei – towards central Asia. All three rivers currently flow into the Kara Sea in Siberia, and thence into the Arctic Ocean. A major new reservoir in Tobol'sk would have fed the new waterway to central Asia. This would have been more than 12 metres deep and 100 metres wide, and 2200 megawatts of power would have been needed to pump the water up over the continental divide which lies 113 metres above the point where the Ob and the Irtysh join.

The scheme aroused much controversy, not least because it would have reduced the fresh water flowing into the Arctic Ocean by some eight per cent. Critics claimed that this could upset the salinity and/or the thermal balance in the Arctic, resulting in major changes in ice build-up. Scientists were divided, however, as to whether the changes would produce more or less ice, and as to what effects elsewhere either change would have. Perhaps it was the scale of the uncertainties, rather than the objections themselves, that led to the cancellation of the Siberian exercise.

Although the Siberian scheme was never to have attempted to refill the Aral Sea – Soviet planners decided long ago to let the sea die in favour of increased irrigation in Central Asia – it was intended to halt the emptying of the even bigger Caspian Sea. Nature, however, seems to have worked some peculiar tricks here. Although the level of the Caspian declined all through the 1970s, the sea has since begun to refill. It may be that this was one reason why the Siberian scheme was cancelled; certainly, the increased flow to the south from Siberian water would have added to problems in the Caspian where in the 1980s rising water levels were threatening port facilities and the Buzachi Peninsula's oil fields.[2]

While the Siberian plan can be forgotten for the time being, plans to divert water from the north and west of the country to feed the Kama and Volga Rivers, and hence the Caspian Sea, are forging ahead. The first phase involves the annual transfer of some 19 cubic kilometres of water from a number of lakes and rivers in the north of the country, including Lake Onega and the Upper Pechora River. Work has started and is due for completion by the end of the century. It will improve navigation on the Kara and Volga, provide considerable additional hydropower and improve irrigation over about four million hectares. The first phase of the scheme involves the creation of some 2271 square kilometres of reservoirs.[3] Two further phases are planned, the second of which involves building a barrage in the White Sea to create a freshwater reservoir there. However, one of the aims of these schemes was to boost water levels in the Caspian Sea, and it is no longer clear what is happening to the Caspian. The implications for the Soviet Union's major river diversification plans have yet to be reassessed.[4]

China is also involved in the large-scale transfer of water resources from one region to another. Some 5 per cent of the Yangtse River is to be diverted to China's semi-arid northern provinces by a 1000 kilometre canal. This canal will be based on the Grand Canal which was rebuilt by Kublai Khan in 1291 and which runs from west of Shanghai to Beijing. The diversion will cross under the Yellow River and, when completed, will deliver 47 cubic kilometres of water a year.[5]

Doubtless, irrigation water will continue to be transferred from water-rich basins to water-poor ones in the future. But the heyday of these most grandiose of all engineering schemes may have already passed. As the World Resources Institute puts it, "In the United States, the era of building huge federal dams and long aqueducts and canals appears to have ended. The scarcity of good dam sites, high costs, food surpluses, and environmental opposition explain why."

Desalting the Sea and Melting Icebergs

There are, at least in theory, other ways of getting water on a large, or largish, scale. In the early days of nuclear power, when optimists believed that the time of everlasting cheap energy had finally arrived, it was thought that all freshwater problems would be solved by desalination plants. As the technocrats talked of such utopian ideals as "the final conquest of disease" and the "greening of the desert", they dreamed up plans for enormous agro-industrial complexes in places such as Israel's Negev Desert, which would be powered by nuclear reactors and watered from desalinated sea water.

The cheap nuclear power never materialized, of course, and it turns out that the business of extracting salt from sea water is actually very expensive. This means that its use is ruled out for large-scale applications such as irrigation.

But desalination is useful for supplying domestic and even industrial water to those who can pay for it, mostly in arid or semi-arid regions where other supplies are very limited. Though expensive, desalinated water is often cheaper than water hauled over long sea routes by tanker. (Indeed, the first desalination plant was built by the British in Aden in 1869 specifically to provide fresh water for ships' boilers for vessels calling at the port.) Desalination plants currently provide up to 10 million cubic metres of fresh water a day, much of it to island inhabitants, for example in Bermuda, Malta (where 10 per cent of supply is desalinated), the Greek islands and the near-island of Gibraltar (where rainwater is also collected from every roof). Desalination plants are also in use in Belgium (which has a 40,000 m³/day plant at Ostend), Egypt, Israel, Italy, Mexico, the Netherlands, Spain, Saudi Arabia, the United States and the Soviet Union.[6]

The major large-scale use of desalinated water is the United States' supply to Mexico. In 1944 the United States signed a treaty with Mexico to to provide that country with set annual volumes of water for irrigation via the Colorado River. Colorado water is used extensively for irrigation in the upper reaches of the basin. Every effort is made, of course, to return the used water to the river so that it can be reused downstream. But the salts washed out of irrigated fields upstream cause increasing problems. It is estimated that 84 per cent of this salt finds its way back into the river.[7] As a result salinity levels in the river rise from 50 to nearly 900 milligrams per litre at the Mexican border. In Mexico some 200,000 hectares of land depend on Colorado water for irrigation. Repeated complaints from Mexico have led the United States to install a desalination plant in California to reduce salt levels in the water flowing out of the United States and into Mexico.

In theory, at least, it may in fact be cheaper to tow icebergs from the Antarctic and melt them down for tap water than it is to desalinate salt water. Iceberg farming, it seems, could make sense, even though about half the iceberg would melt before it could be towed far enough to eke out water supplies in an arid area.

Studies were made of the costs of using icebergs from the Antarctic Amery Shelf to provide water in Australia, and from the Ross shelf to provide water for South America's Atacama Desert. In 1973, when the initial studies were done, it was calculated that iceberg water could be 100 times cheaper than water delivered from a large desalination plant.[8] Even so, no one has yet towed an iceberg home.

Nor have scientists yet managed to implement reliably another of their favourite fantasies: weather modification, principally through cloud seeding. A few isolated experiments do seem to have resulted in short, sharp showers that would not normally be expected. But the effects are difficult to reproduce and almost impossible to predict. In the environmentally conscious 1990s, experiments to change the weather are likely to receive short shrift from those who make it their job to keep the planet's environment as intact as possible.

Grandiose plans for water diversion and use of water on a large scale are not restricted to the developed countries. Both China and India have undertaken major hydraulic projects for re-routing substantial quantities of water. Libya has embarked on an ambitious project to extract water from an aquifer lying under the Sahara to provide irrigation for a quarter of a million hectares of cropland near the Mediterranean at a total cost estimated at $25,000 million.

For something like six millennia, the human race has been involved in a battle to control water resources. In one sense, victory has already been achieved: at least a third of the world's run-off has been tamed and regulated. Large dams restrict the chaos of floods, and hold back the flood water to serve for other times in the year when the rains may fail. The immediate purpose of this exercise is to increase the proportion of the world's run-off which is not lost in floods – the stable and reliable run-off.

In the inhabited areas of the earth, most of the major rivers have already been at least partially developed in this way. But the need for water is now so great that plans are being extended higher up the river reaches, and into remote areas where access is difficult. The costs mount. And, all too often, those involved with taming the muddy and irregular torrents forget that their original purpose was not simply the provision of a dam but the development of the entire watershed for the specific purpose, as President Roosevelt once put it, of "reclaiming land and human beings".

The Tennessee Valley Authority

There is one sound model for the development of large-scale plans for the management of river basins: the Tennessee Valley Authority (TVA) created more than 50 years ago in an attempt to improve the livelihoods of the thousands of rural poor who scratched their livings from along the banks of the Tennessee River in the United States.

Over its long history the TVA has brought wealth to many of its inhabitants, provided cheap electrical power, catalysed industrial development, controlled floods, increased navigability, and reforested the watershed's slopes – and all this in an area that was described in the 1930s as an "economic and environmental disaster area".

From the start, it was laid down that construction work should be carried out by indigenous labour – and thus began the slow process of economic recovery. Rural poverty was alleviated by the generous welfare and educational programmes that were made available to TVA employees. Over the years, the programme has been deliberately spread away from the Tennessee itself, and out along its tributaries. The people affected have been able to tailor schemes to their individual requirements, and bottom-up planning has been intrinsic to the scheme's success.

So successful, in fact, was this whole approach to river basin development that TVA engineers and planners have gone out into the world as the "modern missionaries" of watershed development. Via international organizations such as the World Bank, they have exported the concept of watershed planning to the Far East, India and Pakistan, the Middle East, Africa and South America. The results have been, generally, as dramatically poor as they were good in their home state of Tennessee.

Before asking why, the TVA today merits a few provisos. The scheme was borne before the greening of America, and before the environmental movement of the 1970s and 1980s got underway. Some of what were regarded as signal achievements in the 1930s, 1940s and 1950s now look less attractive. The river itself could not provide all the power for the developments the TVA fostered. So coal-fired and nuclear plants were added to the TVA generating capacity. Complaints from Canada about acid rain were met by a programme that reduced sulphur dioxide emissions by half – but, of course, still failed to eliminate them.

It is charged that today the TVA has become nothing more than a giant electric power utility which produces cheap electricity and encourages extravagant use. Indeed, TVA residents do use substantially more electricity than US citizens elsewhere. And there are charges that the TVA now exists to help big business and the State, and that the poor and needy are hard done by. The TVA, it is said, no longer serves its original purpose of reclaiming land and human beings. With few exceptions, the TVA model has exported badly.

Damming the Narmada River

The Narmada Valley in the Indian state of Gujarat is the scene of intense activity. Engineers there are rushing to complete the giant Sardar Sarovar Dam that they believe will bring an end to Gujarat's problems of recurring drought.

The dam will be 1210 metres long and 139 metres high. From the reservoir it impounds will lead the largest canal in the world – 750 metres wide and 445 kilometres long, reaching from south Gujarat into Rajasthan. On its way it is planned to irrigate 1.8 million hectares, supply 3.5 million cubic metres of drinking water and produce 1450 megawatts of electricity. The main canal will feed 37 other, smaller canals. They, in their turn, will feed yet smaller ones. In all, 72,000 kilometres of new waterways are to be built. The cost will be more than £3500 million, the project will take 22 years to complete, and according to its builders should serve the community for at least a century.

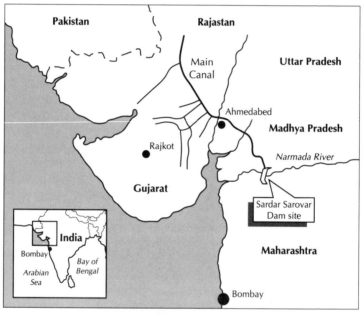

Sardar Sarovar dam and the Narmada river

The canal itself will be lined to prevent seepage and the project will be the third in the world to use computers to open and control flood gates and sluices at 10 to 15 kilometre intervals, and thus regulate the flow of water downstream by remote control. The surrounding water table will be watched through piezometers located in nearby wells. Legislation will require farmers to abide by a water quota from the canals, which will depend on the crops they grow and the amount of ground water they can extract, so that water is not needlessly squandered. More than 40 million trees will be planted to compensate for the large areas of forest that the project will destroy.

Early in 1989 the popularity of this scheme was attested by 10,000 people who marched to the site of the dam to protest against its construction. Hundreds lay down on the road to stop vehicles reaching the site. Doubtless most of them lived in the areas to be flooded, which will force 70,000 people to move and destroy nearly 14,000 hectares of forest land. Doubtless many of them were also protesting against the malaria and other water-related diseases which the project is sure to spread in its wake. And against the destruction of important shrines and temples by the flood.

ANNUAL SEDIMENTATION RATES IN INDIAN RESERVOIRS
(million cubic metres)

	Assumed rate	Observed rate
Bhakra	28.36	41.27
Maithon	0.84	7.37
Mavurakshi	0.66	2.47
Nizamagar	0.65	10.76
Panchet	2.44	11.75
Ramganga	1.34	5.38
Tungabhadra	12.08	50.62
Ukai	9.18	26.83

Source: Centre for Science and Environment, *The State of India's Environment, 1982*. New Delhi, Centre for Science and Environment, 1982.

But they had many other grounds for concern. How soon will the project lead to waterlogging and salinity? What effect will the dam have on fisheries downstream and in the Narmada estuary? Who will look after all the new trees once they have been planted? Most important of all, for how long will the scheme work, assuming it does work?

Experience in other parts of India is not encouraging. Sedimentation rates have often turned out to be several times (in one case, more than sixteen times) those predicted from theoretical studies (see table on page 121). And there is little evidence to suggest that the Sardar Sarovar reservoir will prove an exception.

Enormous though the Sardar Sarovar project is, it is but part of an even more massive scheme, India's – and maybe the world's – largest, known as the Narmada Valley Development Project. This involves no fewer than 30 major dams, 125 medium ones and 3000 minor ones along the Narmada's 1300 kilometre length. It is estimated that the entire project would irrigate 4.8 million hectares and force more than a million people to move home and farm. Even the World Bank does not recognize the project.[9]

Considering India's troubled history of failure with large dams and irrigation projects, and considering levels of current protest about even small dams, it is unlikely that the full project will ever be finished. India's prime minister, Rajiv Ghandi, certainly didn't mince his words in 1986 when he addressed a meeting of state ministers of irrigation:

> The situation today is that, since 1951, 246 big surface irrigation projects have been initiated. Only 65 of these have been completed and 181 are still under construction. We need some definite thrusts from the projects that we started after 1970. Perhaps we can safely say that almost no benefit has come to the people from these projects. For 16 years we have poured money out. The people have got nothing back, no irrigation, no water, no increase in productivity, no help in their daily life.

What they have got, which Ghandi did not mention, is a fair degree of social chaos and an increasing incidence of water-borne disease. While families in developed countries can move from one region to another, even one country to another in Europe, without major disruption, the same is far from true in the developing world. Most of those affected by large dams are rural dwellers, living off the land. They have practised the form of farming suited to their lands often for as long as their history goes back. They have their own culture, their own tribes, their own religious convictions. Uprooting such people, and moving them to land they have never seen, is bad for agriculture and worse for social cohesion.

Moving People

Building large dams almost always means moving people. Relocation is never popular, and it has been carried out on a massive scale in Asia, Africa and Latin America. When the Kariba Dam was built on the Zambesi River, 57,000 members of the Tonga tribe had to be moved. For a people who had lived in the area for as long as their history, the cultural shock was inevitable – particularly since the tribe had to change to dryland farming techniques when they were used to systems based on river flooding.

The food shortages and disease that followed in the wake of the relocation, however, were not inevitable, and resulted from poor planning.

Not that planning resettlement on this scale is easy. Something close to social chaos followed the even bigger resettlement that accompanied the Volta River project in Ghana, which involved the creation of the 8482 square kilometre Lake Volta. This was not a multi-purpose scheme but one specifically designed to provide hydropower, which it was hoped would encourage industrial development and economic growth. Neither materialized. Although jobs were created in both the bauxite and the electricity generating industries, the cost per work place was astronomical, double or even quadrupal the cost of creating jobs in the United States at the time. According to one observer, ". . . from the very beginning, the Volta scheme was conceived, constructed and finally coopted by the international aluminium companies and the international banks for their own profit maximization."[10] But there were other reasons as well. The Volta River project was planned on the basis of data collected during the 1950s and 1960s alone – earlier records were quite inadequate. The 1970s and 1980s, in comparison, were much drier, and the flow rates and hydropower production anticipated in preliminary studies were never achieved in pactice.[11]

But 78,000 people – some one per cent of the country's population – had to be moved from 700 small towns and villages to 52 different resettlement locations.[12] Residents were able to choose between resettlement and compensation for their land, and 85 per cent chose to move. They were supplied with a roofed space containing one room, and were invited to construct a further three rooms under the roof in any way they wanted.

It was intended that generous areas of land be provided for four kinds of farming: arable farming (4.9 hectares), tree crops (2 hectares), the raising of intensive livestock (1.2 hectares) and pastoral farming (12.1 hectares). Tractors and cultivators would also be provided. But advice on how to operate the new equipment and the farming systems they were intended to serve was woefully lacking, and the scheme neglected to provide markets for the produce.

As the failures of this ambitious programme became evident, the authorities decided to settle for less. They provided each family a mere 1.2 hectares of uncleared land on which to practise subsistence agriculture. The response was emphatic: 60 per cent, some 42,000 people, moved out.

Even without food shortages and disease, relocation schemes that are poorly planned – and most have been – can cause immense suffering. Writes Lloyd Timberlake:

> The Talata-Mafara Dam and irrigation project in northern Nigeria involved moving 60,000 peasants during three years of construction work, with no compensation. As they could earn no money during the period, many had to mortgage their land to the bankers, civil servants and businessmen of Kano. The elite of Kano also benefited from the Kadawa irrigation project, as farmers had to rent their lands to those who had the money necessary to invest in irrigation.[13]

There are no global estimates of how many lives have been dislocated by resettlement programmes over the past few decades. In Thailand 11 hydropower projects have forced more than 23,000 households – containing an estimated 140,000 people – to move.[14] More than 1500 large dams have been erected in India over the past three decades, and the authorities have long since lost count of both the number of people resettled, and the economic and human costs involved. The Aswan High Dam in Egypt forced the resettlement of 120,000 people – and some of the country's most precious ancient monuments. In Nigeria, the construction of Lake Kainji forced 50,000 to move.

Relocation on this scale today might well prove impossible. Over the past decades, people have learned better how to exercise their power, and are less easily forced to move from one place to another. As word gets round that alternatives to large-scale projects are possible, that the Chinese system of chains of small dams works well, and that smaller systems can be better controlled and provide more advantages, resistance to the gargantuan schemes of the past – and the future – will swell.

It needs to. Despite all the evidence, more large-scale water atrocities are being built and are in the development pipeline. Senegal, Mali and Mauritania have joined together to plan major developments of the Senegal River, including two large new dams. And while it is true that the new international body charged with planning the development has given extensive thought to problems of resettlement, and implementing the irrigated farming that will result, past experience does not augur well.

From the Upper Mekong Delta in Thailand, to the Ganges in India and the Nile in Egypt, the experience has been much the same. Big dam projects have brought wealth to a few, kudos to some politicians, and despair to millions. They have also brought disease.

The Impact of Disease

Water carries disease. In fact, it carries almost 80 per cent of disease throughout the world. Half the world population is estimated to suffer from infections caused by water-related disease.[15] And when large new volumes of water are created, the risk of disease in tropical countries rises sharply. The diseases associated with dams built to provide irrigation water and hydropower include schistosomiasis (or bilharzia), yellow fever, malaria, river blindness (onchocerciasis) and liverfluke infections.[16] Malaria was the cause of early failures to construct the Panama canal, when 57 per cent of the 86,800 workforce were incapacitated from the disease and 5627 people died from it.

Schistosomiasis is often the first to appear. Caused by a flatworm carried by a water snail, this disease affects as many as 300 million people in developing countries, leading to debilitation and vulnerability to other diseases. Its spread and severity often follow hard on the heels of the introduction of irrigation schemes.

A study in Egypt of four selected areas showed that the incidence of the disease climbed from 10 to 44, from 7 to 50, from 11 to 64 and from 2 to 75 per cent of the population within three years of the introduction of a new irrigation scheme. A survey after the construction of the dam to create Lake Volta showed that the incidence of infection with schistosomiasis among children under 16 in the resettled areas rose from 3 to 37 per cent in just one year.

These figures are not exceptional. On the contrary, they are typical of what happens when large new areas of static or sluggishly flowing water are introduced into hot climates. The fact that these projects continue, in spite of the evidence against them, has much to do with greed and big money. Top-down planning, with ritual genuflections towards grass roots movements, is not enough. An Indian writer, B. Dogra, claims bluntly

> . . . experience with large dams has been disastrous. Yet that experience is ignored by the government, which continues to approve new dam projects despite the lessons of the past.[17]

Hydropower and the Environment

The environmental costs of the dams and reservoirs needed to bring irrigation to dry lands have been described in Chapter 4. Often, these are multipronged projects designed not only to supply irrigation water but also to provide hydropower and to stimulate industrial and economic development. Often they fail to do so, and many end up – as the critics of the TVA allege – as schemes that serve mainly to provide electrical power, most of which is used by urban rather than rural populations.

But thousands of projects have also been specifically designed to provide only hydropower. In many of these schemes, scant attention has been paid to the environmental and social costs involved. They are considerable. And they occur in three different areas.

Upstream of the site, people have to be resettled from areas further downstream. And fish can no longer migrate upstream past the dam.

At the site of the dam, people must be relocated, sediments accumulate, the levels of nutrients in the water rise and eutrophication (over-oxidation) can set in. Transport is disrupted and any existing fisheries are either destroyed, or required to adapt to the new species mix that eventually develops.

Downstream there is erosion and soil loss, a decline in nutrient level, often a loss of the annual floods on which existing farming systems were based, and declines in fish levels.

Yet, in developed countries at least, there are signs that common sense might yet prevail. During the 1980s there was fierce debate over plans to develop the lower reaches of the Gordon-Franklin River in Tasmania. It was charged that the development would destroy 30 per cent of the remaining pine stands in the region, flood large parts of important wilderness and recreational areas, lead to severe soil erosion, alter the ecology of the area and encourage invasion by different species of plants and animals, destroy important archaeological and geological sites, and lead to severely increased fire risks.

In 1983 the Australian Federal government passed legislation that required all work on the project to stop. The government paid out more than $276 million as compensation to the Tasmanian administration to cover losses, provide alternative employment and encourage the development of hydropower sites in other regions.[18]

The Australian government, in this instance at least, has taken the line advocated in the quotation as the beginning of this chapter. Unhappily, there are still few signs of such enlightened actions in developing countries – even though, as chapter 10 shows, there exist workable, environmentally benign and socially acceptable alternatives that could be sustainable far into the future.

Notes

1 Sattaur, Omar "India's Troubled Waters", *New Scientist*, 27 May 1989.
2 World Resources Institute and International Institute for Environment and Development, in collaboration with the United Nations Environment Programme, *World Resources 1988–89* (New York: Basic Books, 1988).
3 P.P. Micklin. "Soviet River Diversion Plans: their possible environmental impact", in E. Goldsmith and N. Hildyard (eds.), *The Social and Environmental Effects of Large Dams* (Cornwall, UK: Wadebridge Ecological Centre, 1986).
4 World Resources Institute, *et al.*, *op cit.*
5 World Resources Institute, *et al.*, *op cit.*
6 UN Economic Commission for Europe. *Long-term perspectives for water use and supply in the ECE region* (New York: United Nations, 1981, Sales No. E.81.II.E.22).
7 Law J.P. and A.G. Hornsby, "The Colorado River Salinity Problem", in G.V. Skogerboe (ed.), *Water and Energy Development in an Arid Environment: the Colorado River Basin. Water Supply and Management* 6(1/2), 87–103.
8 Weeks W.F. and W.J. Campbell, "Icebergs as a Freshwater Source:

an appraisal", *Journal of Glaciology*, 12(65), 207–33, 1973.

9 Sattaur, Omar, "India's Troubled Waters", *op cit.*

10 Graham R. "Ghana's Volta Resettlement Scheme", in E. Goldsmith and N. Hildyard (eds.), *The Social and Environmental Effects of Large Dams, op. cit.*

11 Falkenmark, Malin, personal communication.

12 Biswas, Asit K. "Environmental Implications of Water Development for Developing Countries" in Carl Widstrand (ed.), *Water and Society: Conflicts in Development* (Oxford: Pergamon Press, 1978).

13 Timberlake, Lloyd, *Africa in Crisis: the causes, the cures of environmental bankruptcy* (London: Earthscan Publications, 1985).

14 R.P. Lightfoot. "Problems of Resettlement in the Development of River Basins in Thailand", in S.K. Saha and C.J. Barrow (eds.) *River Basin Planning* (Chichester: Wiley, 1981).

15 Barabas, S. "Monitoring Natural Waters for Drinking Water Quality", *World Health Statistics Quarterly*, 39(1), 32–45, 1986.

16 Biswas, *op.cit.*

17 Dogra, B. "The Indian Experience with Large Dams", in E. Goldsmith and N. Hildyard, *op. cit.*

18 McDonald, Adrian T. and David Kay, *Water Resources: issues and strategies* (London: Longman Scientific and Technical, London, 1988).

Chapter 9

Traditional Solutions

*It is futile to expect magic solutions to the problems of
semi-arid regions.*

Norman W, Hudson
Soil and Water Conservation in Semi-Arid Areas, *FAO*

In 1730, in Rajasthan's northern Tar Desert, 363 members of
the Bishnoi community were beheaded in a fierce struggle
to stop outsiders felling their trees for fuelwood. In January
1988 the Indian government commemorated the event by
naming the Bishnoi village of Khejare as the first National
Environment Memorial. The government has thus given official
recognition to a remarkable people whose culture and religion
has enabled them to withstand centuries of drought – and, more
particularly, the recent five-year drought in which average
annual rainfall in the Tar Desert fell from 250–600 millimetres
to only 10 millimetres.

The story correctly begins in 1452, with the birth of Jangesh-
war Bagwhan, or Jamboje, the Bishnoi saint who linked the
principles of ecology and the Bishnoi way of life. His book of
revelations *Jamsagar* ("show the people light") lays down basic
rules that have preserved the Bishnoi environment and their
way of life ever since.

The Bishnoi will not kill animals or trees, will not eat
after dark for fear of accidentally eating (and thus killing)
insects in their food, and revere the khejare tree, *Prosopis
cineraria*, whose leaves provide their cattle with a food – it
has since been discovered – that contains 11 to 14 per cent
protein. The tree is carefully pruned so that cattle can eat the

lower leaves while the upper branches are used for fuel and building poles.

The Bishnoi keep neither sheep nor goats for fear of ruining their land but rear an average of four cattle per head. They are settled pastoralists, living in groups of about a hundred, who eat mainly milk, yoghurt and cheese, and use dried dung for cooking (fearing that its acidity would harm the roots of their seedlings were it applied as fertilizer).

They also gather berries, leaves and fruit from surrounding bushes and trees. Water collected in cisterns enables them to grow two varieties of millet in summer, and radishes, carrots, onions and garlic in winter. Dairy products are traded in nearby towns for oil, sorghum, wheat and lentils.

Remarkably, this simple code of life appears to have enabled the Bishnoi to survive the recent drought without any apparent impact. Research by the Indian society for environmental protection shows that the drought has hardly altered the incidence of disease and mortality among the Bishnoi.

Michael Tobias, who recently visited the Bishnoi, and on whose description this account is based, watched a group of Bishnoi laboriously digging a well by hand for water they knew might lie 80 metres below the ground. Nevertheless, they were in cheerful spirits, and stopped work to welcome him and discuss their project. He writes:

> Passing by the circular mud houses of their village, which rest cosily on earth that has baked for too long, I was still not sure whether I was seeing a typical scene of survival such as the desert must always impose, or a people in terrible turmoil from the drought. But these are the Bishnoi. Their love of the earth manifests itself through pragmatic beliefs and strategies, deeply considered, passed down from generation to generation, and not easily overturned. They will survive these hard times.[1]

The Bishnoi's environmental advisor, Professor S.M. Mohnot, of Jodhpur University, believes that there are lessons to be learned from the million-strong Bishnoi. "Our films, books, lectures and endless symposia are useful to a degree", he says. "But ultimately they miss the boat. They come from the city, and rarely filter down to the villages in crisis. What we need to see happening is a reversal – an ecological sensibility that *starts* at the village level. *That* is why the Bishnoi are so significant."

Run-off Farming

There are few drier places than the Negev Desert, with its annual rainfall of around 100 millimetres. Yet archaeological research – and, indeed, inspection with the naked eye, for the remains are still visible in many places – has revealed that the desert once supported many thousands of small farms consisting of terraced fields of 0.5 to 2 hectares. Crops flourished there, unaided by sophisticated pumps or underground aquifers.

The only water available – as today – were the short, gentle winter rains that characterize the area. This run-off was collected via a series of intricate water distribution systems, which led the water from one terrace to the next. Each field needed to use the water falling on a catchment area 20 to 30 times larger. It was led from there to the fields by small ridges of earth and stone built diagonally across the valley slopes, which channelled the water down the hillside to an entry on the field. Crops were planted after the first winter rains had infiltrated the soil, and they continued to receive water from the system with each new rain. However, the systems were so efficient that even a couple of rains after planting were sufficient to raise a crop. Because winter rains in the Negev can occasionally be heavy, most of the terraced fields contained a spillway to allow excess water to run-off and thus limit the extent of flooding.

In some areas of the Negev, whole valleys were devoted to farming. Water was taken from the stream or river that ran down the valley during the wet season, and small dams, five or six metres high, were used to divert the water into canals which fed the farms. The most well known of these, the Marib

Dam, was 14 metres high. From it water was led via small canals, four to eight metres wide and up to three kilometres long, to the terraced fields.

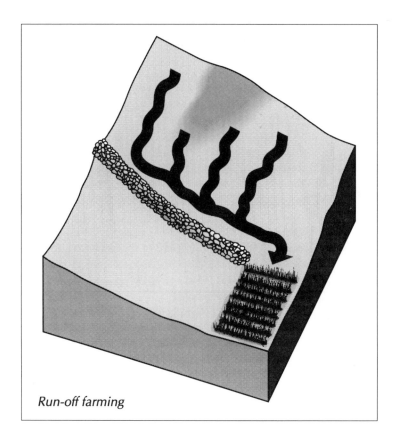

Run-off farming

Recent research has shown that these ancient systems were more sophisticated than is immediately apparent. One of their curious features is that they are characterized by small mounds of stones. Great effort apparently went into collecting up all the loose surface stones in the catchment area, and heaping them into mounds. Common sense suggests this would decrease run-off, since loose surface stone ought to prevent water seeping into the ground before it reached the fields. Common sense, it seems, is wrong in this instance. Experiments have shown that removing the stone does actually increase run-off significantly, for a number of quite complicated hydrological reasons. Among the many factors involved is that stones slow down the rate of run-off, and thus increase evaporation. Furthermore, since the soil round a stone remains wetter for longer, wild plants have a greater chance of surviving there. Plants and trees, as we have seen, are an effective means of absorbing run-off. On a catchment area, however, it pays to remove all vegetation to allow the water to run down the slope. Research in the northern Negev shows that removing vegetation decreases the rate at which water infiltrates the soil by 40 to 50 per cent and increases run-off by 7 to 21 per cent.[2]

At one time, it was thought that these stone mounds might have been built as dew collectors. Dew would drip down the stones and be run-off to water a nearby tree. Research shows that not enough dew is formed to drip down the stones, however. But dew may nevertheless play a role in the ancient agriculture of the Negev. At Avdat, for example, the annual dew fall is 25 to 35 millimetres a year, compared to rainfall of 150 millimetres. While the amount of dew is insufficient to provide any water by run-off, what falls on the crop may be significant; in 1962–63, for example there were 28.4 millimetres of dew compared to only 25.6 millimetres of rain.[3]

Size is critical to the ancient Negev agriculture. For example, it would not be possible to replicate this system on a larger scale, with fields of say 100 hectares fed by catchment areas of 2000 hectares or more. One reason is that large catchments always contain more depressions, and places where water can leak away, than small ones – and on small ones depressions can be filled and leaks plugged. Even more important is the amount of time needed for the run-off actually to happen: a large roof takes much longer to drip into its gutter than a small one. So with catchments. This means that the crop in the hectare-sized fields of the ancient Negev would benefit from even the lightest rains, while those waiting for water from much larger catchments would never receive any. In this, small is certainly more efficient.

The Negev is not the only place where rainwater was harvested to grow crops. Similar systems existed in many areas of the Near and Middle East, in Pakistan and northern India, in the Soviet Union, in Africa and in the United States.

In many of these systems, water control was not as elaborate as in the Negev; in some, it did not exist at all, in that natural features were used to provide both the catchment area and the channelling of the water towards the fields. For example, the Navajo in Arizona used run-off from rocky outcrops on hills to channel rainwater down to their fields below, where they grew maize, squash and melons. Sometimes small dams were erected to hold the water in the fields for longer. The Hopi Indians used flood waters to saturate their fields before planting. Three-quarters of their agriculture was based on this technique – though much of it is no longer used.[4] And the Papago Indians of southern Arizona used fences made from posts and brushwood to spread floods over their fields. The fences also trap nutrient-rich silt, and thus play a double role in both water harvesting and nutrient harvesting.

Another common technique was to plant crops in the silt left behind when rivers dried up – a method which utilizes both the water left behind as well as the nutrients. In Ethiopia's Woito and Lower Omo valleys, for example, the flood plains are cleared and planted to maize and sorghum when the annual flood recedes. Similar systems are used in central Asia (where they are known as *kair* farming) and in Turkmenistan, where they are known as *khaki* farming. Tunisians used a system of *gessours*, in which they planted rows of olive trees across the stream bed. A small dam upstream was used to slow down the flood; it also caused the silt to build up and thus produce a system of terraces. Similar systems were used in both China and Mexico.

Northern Pakistan boasts one of the more amazing water collection systems, based on drainage channels, called khuls, that were hewn from the steep rock faces of the mountainside. A metre or so deep, these canals collected water from melting glaciers and snow from large areas of this mountainous region, leading it down to the rich alluvial soils at the bottom. There all manner of crops were grown – and still are – in the oases of this cold and otherwise intensely dry mountain desert.

Flooding the Fields

The ancient systems described so far all used rain or flood water to water their crops. What they have in common is a method of storage that wets but does not saturate the soil, usually on several occasions during the growing season. Where rainfall is very seasonal, methods have been developed that rely on a once-only water delivery system. In this case, the water is used to saturate the soil, often to considerable depth. As the flood recedes, the crop is planted and matures almost entirely by drawing on this water; with luck, a little extra rain helps but many crops have to survive without. In the Sudan, this system is used to grow a form of millet that matures in only 80 days. The soil is first flooded by rainwater collected from gently sloping catchment areas known as "teras".[5]

Similar but much more elaborate systems have existed in the semi-arid areas of Pakistan and India for more than five centuries. Long low banks of earth are constructed to hold back the monsoon floods, and submerge the fields to a depth of a metre or more. The crop is planted only after the flood waters have been drained off. In Rajasthan the flooded areas are called *khadin* and in other areas of India, such as Bitar and Uttar Pradesh, they are called *ahars*.[6]

Many of these systems are built on a grand scale, and cover vast areas. There are an estimated 800,000 hectares in Bihar alone. In Bihar they occur on very gently sloping land, with the result that an area extending several kilometres back from the bank can be flooded to a depth of a metre or more. In Rajasthan, where rainfall is lower, the *khadin* occur near hills and ridges, which act as catchment areas, and are usually on steeper land.

When the rains end, gates in the banks are opened, and the flood water is then often used to flood another *ahar* downhill. The banks are usually not more than three metres high, though they may be several kilometres long. Winter wheat or pulses are then grown during the five-month long winter season, when evaporation rates are at their lowest. A rice crop can also be grown during the summer while the land is still flooded.

These systems integrate a number of forms of agriculture in subtle ways. One of their advantages, for example, is that the soil is regularly washed by the flood, and salinization kept at bay. Furthermore, the earth banks act as silt traps, with the result that the arable land is gradually levelled, and its sandy soils slowly replaced by more clayey ones, which hold water better. Thus the performance of these systems tends to improve with age. The nutrients that are washed into the fields every year also boost yield.

The water trapped in this way is put to several uses. While it is building up, and before it is time to plant, some of it may be tapped off and led to other fields where irrigation is practised. One *khadin* near Bharatpur which was built in 1895 has a retaining bank 19 kilometres long which is used to flood 4100 hectares. But it also provides irrigation water for another 4800 hectares.

In many places, and particularly in Rajasthan's Tar Desert, the effects of this system of farming are also exploited by livestock. One of the effects of a flooded *khadin* is a raised water table. Many pastoralists exploit this by sinking wells downhill of the earth bank to provide water for their livestock. These wells have beneficial effects for the arable farmers above. By increasing the drainage rate from the flooded areas, they help reduce salinization.

The *amar* and *khadin* are among the most impressive of

all traditional dryland farming systems. Their popularity is
attested by their size. But though many are built and operated
on a large scale, the individual units within them do not follow
the same pattern. A 150 hectare *khadin*, for example, may be
worked by as many as thirty people, some with a holding of
little more than one hectare. These systems are therefore also
impressive in the high degree of social organization that their
efficient operation entails.

It is no accident that Africa has hardly been mentioned in
this chapter. The literature on water harvesting is short on
African examples. This may be because there are few; or it
may be because those who have studied dryland African
agriculture have taken little notice of simple, apparently
primitive systems designed to grow only a few sacks of such
African staples as sorghum and millet.

But there is one important exception, which involves sorghum growing in Kenya's southern Turkana region, where
rainfall is usually less than 200 millimetres a year.[7] The
crops are grown only by women, each working a garden of
about one-tenth of a hectare. The sorghum is grown in various
areas in the region of the Kerio River, and many of the water
harvesting techniques already described on other continents
are used. Like the Navajo systems, water is collected from rock
outcrops and used to irrigate the fields. Flash floods are used
to provide water in other places. And, as in Tunisia, China
and Mexico, crops are also planted on stream beds and along
river banks.

This traditional system is potentially important for a number
of reasons. One is that it uses a very quick-growing form of
sorghum, which matures in only 65 days. This variety is
now being tested elsewhere. Another is that it does indicate
remarkable adaptivity to different conditions, all within a
small area. And a third is that it exists. And if it exists in
southern Turkana, it – or similar systems – may well exist
elsewhere.

Water harvesting in Africa is not, of course, a completely
closed book. A few basic techniques have been described from
most areas of the continent. For example, stone barriers are

erected across stream beds in Somalia to divert water to fields, stone bunds are used to control run-off in countries such as Burkina Faso and Niger, and sorghum is grown behind small dams in Mauritania.

In some places, more elaborate conservation systems have been developed, and are still used. The Konso tribe, in Ethiopia's central highland plateau, for example, employ an ingenious farming system which depends essentially on piling up the rubbish from last year's crop in the right place. This simple, if untidy, technique enables the Konso to grow crops and trees on quite steep land; run-off is controlled by the rows of mixed earth, old maize stalks and other agricultural waste which are piled up along the contours, forming miniature terraces.

In the Yatenga region of Burkina Faso one traditional technique of growing cereals is to dig small pits or depressions in the field, each about 20 centimetres in diameter and about one metre apart. Manure is placed in the bottoms of these pits, which also trap leaves blowing across the ground surface, which then act as a mulch. Termites apparently then add to the process by tunnelling their way up from underneath to feed on the manure and leaves in the pit. The termites help aerate the soil under the pit and provide tunnels which help root growth and water infiltration.[8]

Small-scale Irrigation

Probably the most exquisitely developed traditional irrigation system in the world is that of the Balinese. Not that the system is technically advanced; on the contrary, it relies mainly on loose stone dams and weirs, and on delivering water to the fields via the hollowed-out trunks of coconut trees. But what is advanced are the elaborate social mechanisms that have been evolved to ensure the system works, and that water is distributed equitably to its users, according to how much land they farm.

Balinese success in this area is not unique. In fact, the

contemporary critics of irrigation projects who argue that irrigation never works properly because it is impossible to devise equitable ways of regulating them, need to do some serious historical research.

It is doubtful, for example, that the Hohokam Indians in North America's Sonoran Desert would have gone to the trouble they did to irrigate had they not also evolved a social system to make it work. We know nothing of it – except that it worked well enough to merit the construction of 500 kilometres of major canals and 1600 kilometres of minor ones. Their system was constructed a thousand years ago, and irrigated two crops a year.

In Tanzania, the Sonjo and the Chagga tribes grow their crops on the sides of Mount Kilimanjaro. Both use irrigation, and both reflect the importance of their water control systems in their religions.

The Sonjo cultivate three types of land: *hura* land on the valley floor, which is flooded in the rainy season and irrigated in the dry one; *magare* land on slopes above the river which is divided into plots which are cultivated on alternate years to help control salinity; and *isirene* land on natural terraces downstream which depend on water channelled to them by furrows and hollowed-out logs.[9]

The whole system is carefully regulated to prevent over-irrigation, and its maintenance is regarded as a sacred duty, since Sonjo mythology records that the irrigation systems date from the time when the tribe's villages were first created. A Council of Elders oversees maintenance work, and all able-bodied men are required to help in cleaning out irrigation channels and repairing dams washed away by the flood.

The Chagga system is also based on community maintenance. Farmers wanting to use irrigation water brought down from the mountain in furrows must join an irrigation organization. This requires them to help in maintenance work or pay a fine if they fail to do so. But there are reports that the system is breaking down since piped water has been brought to the area. The furrows are no longer so well maintained, and land shortage has forced people to cultivate ravines down to the rivers, removing trees in the catchment area and causing springs to dry up.

Water Control on the Large Scale

One of the finest examples of early water engineering on a grand scale comes – perhaps appropriately – from China. In 250 BC a regional governor in Szechwan, Li Ping, and his son, Er Wang, devised a scheme to tame the floods of the Min River caused by melting snow in Tibet. More than a thousand kilometres of canals were constructed which provided irrigation for 200,000 hectares. The scheme is still in operation today.[10]

The main diversion canal is separated from the Min River by a structure known as the fish snout, which is regularly damaged by the floods. But every year, when the river is low, the canals are cleaned and the fish snout repaired. Flow in the canals is controlled by simple and temporary structures of sausages of rock held together by twisted bamboos. This primitive, yet appropriate system – for it works – has proved capable of controlling an immensely complex hydrological system for more than two millennia.

In many senses, the Min River control system exemplifies a quality totally lacking in today's grandiose water schemes: the combination of large-scale, overall control with small-scale, simple structures. No one item in the Min River scheme is large; even the fish snout that first divides the river is only tens of metres long. Nor is the technology complex. Since it can be easily repaired, it does not have to built to withstand exceptional conditions. Sedimentation is simply controlled by digging out sections of the canal, and then using the rich nutrients obtained to boost crop production. Such ideas would be totally impractical for a modern reservoir, covering hundreds of square kilometres, and designed specifically never to dry out.

Notes

1 Tobias, Michael, "Desert Survival by the Book", *New Scientist*, 17 December 1988.

2 Tadmor, N. H. and L. Shanan, "Run-off Inducement in an Arid Region by Removal of Vegetation", *Proc. Soil Sci. Soc. Amer*. 33(5), 790–793, 1969; and *Micro-catchment Systems for Arid Zone Development* (Jersalem: Hebrew University and Ministry of Agriculture, 1979).

3 Evanari, M. L. Shanan and N. Tadmor, *The Negev, the Challenge of a Desert* (Cambridge, Massachusetts: Harvard University Press, 1982).

4 UNEP, *Rain and Storm Water Harvesting in Rural Areas* (Dublin: Tycooly, 1982).

5 FAO, *Soil and Water Conservation in Arid Areas* (Rome: FAO, 1987, *FAO Soils Bulletin No. 7*).

6 The following description is taken from Arnold Pacey and Adrian Cullis, *Rainwater Harvesting: the collection of rainfall and runoff in rural areas* (London: Intermediate Technology Publications, 1986).

7 Morgan, W. T. W., "Sorgham Gardens in South Turkana", *Geographical Journal*, 140, 80–93, 1974.

8 Wright, P., *Report on Run-off Farming and Soil Conservation in Yatenga* (Oxford: Oxfam, 1984).

9 Sheridan, David, *Cropland or Wasteland: the problems and promises of irrigation* (London: Earthscan, 1984).

10 McDonald, Adrian T. and David Kay, *Water Resources: issues and strategies*. (London: Longman Scientific and Technical, 1988).

Chapter 10

New Twists to Old Solutions

*Information about existing traditions of run-off farming
is inadequate nearly everywhere.*

Arnold Pace and Adrian Cullis
Waterlines, vol. 4, no. 4, April 1986

This chapter is not going to provide the solutions to the problems of the world's drylands. If it could, there would have been no need to write the book. In its review of the techniques that are available, the UN Food and Agriculture Organization goes straight to the point in its first sentence: "This bulletin does not offer easy solutions to all the problems of soil and water conservation in arid areas," it said. "There is no storehouse of tested methods and techniques which can be taken off the shelf for immediate application."[1]

That said, there are plenty of pointers to what has been done and what might be done. Indigenous peoples, farmers from the colonial era, development consultants and local agricultural agents have tried; in places, they have succeeded, more often they have failed. But put together their work provides, if not "a storehouse of tested techniques", then at least a portfolio of ideas.

Nor will this chapter provide a synoptic reference work for the student of water harvesting. Today many people know that in a few places in Ethiopia, at a cost of 1000 person-hours per hectare, a magnificent range of bench terraces has been hewed out of rocky hillsides. They know that rivers can be dammed, ponds dug and irrigation canals made to lead water to arid places hundreds of kilometres away. They know, too, that there exist highly mechanized methods of moving and shaping soil on a large scale to reduce erosion and maximize water availability. These techniques have their place.

But these are heavy solutions, and they are not ones to which local people turn lightly or easily. By and large, they will be ignored in this chapter in the belief that, in the end, only those whose lives are threatened by water scarcity can provide and implement the solutions to their predicament. So this chapter will be focused on light, small-scale and ingenious solutions – and, above all, on solutions that, because they provide immediate benefits to farmers and villagers, have some chance of self-replication without interference from the outside world.

Such benefits can take many forms. The most obvious, apart from an assured supply of drinking water, is the ability to grow more food on less land using less water. This distances the spectre of famine and puts more money in the farmer's pocket. But rural populations need more than this. They need, above all, security – confidence that they can survive the dry years, that investment in better seeds and more fertilizer will not be wasted in a "harvest of dust", to use the title of UNEP's magnificent video on desertification.

In fact, the concept of water security lies at the root of the problems of the world's drylands. We shall explore it further in the next chapter.

The Third Type of Agriculture

The current predicament stems in part from the way the world's agricultural experts have for so long ignored dryland issues. As Robert Chambers writes,

> Of the three main types of agriculture in the world, two of them – the industrial agriculture of the rich North, and the green revolution agriculture of the more favoured areas of the South – have been served quite well . . . But perhaps as many

as 1.4 billion people in the South (roughly 1 billion in Asia, 300 million in Africa, and 100 million in Latin America), among them many of the very poorest, depend on a third type of agriculture which is more difficult, and where yields have changed little since mid-century.[2]

Agriculture in these areas is complex and full of risk. Farmers have to deal with difficult conditions, have to grow many different crops, must mix arable farming with livestock production and small-scale forestry, and must use their products for many widely different purposes.

Chambers argues that the farmers practising these techniques know more about their problems and their solutions than scientists ever will. For these reasons, Chambers claims, dryland farmers need "not messages but methods, not precepts but principles, not a package of practices but a basket of choice, not a fixed menu *table d'hôte* but a choice, *à la carte.*"

What is there in the basket to choose from?

Winter Rains and Summer Rains

The secrets of ancient farming systems in the Negev Desert may well be the main ingredient. Not that they themselves will solve the water crisis in the world's drylands. True, rainfall in the Negev is lower than in many of the places in Africa, India, China and Latin America where drought and hunger have been common over the past two decades. But the Negev techniques are not *per se* transportable to the tropics.

One reason is that the Negev crops are grown during a season of winter rainfall. Evaporation rates then are often as much as five times lower than in the summer. Crops grown under such conditions need much less water than those that must be grown in seasons of summer rainfall, when evaporation rates are high. In the Sahel, for example, a millet crop using Negev catchment techniques and relying on Negev rainfall levels would wither and die long before it reached maturity.

Another reason is that the Negev techniques work in the Negev precisely because it is so dry. This paradoxical statement holds the key to many development failures. Farmers who must survive year after year on rainfalls of less than 250 millimetres a year *know* they have to catch every last drop of water. And their farming systems are adapted to do so.

But where rainfall is usually enough to grow a crop without using run-off, there is less incentive. Bunds, terraces and dams look like hard work when you have grown a successful crop, raised livestock and fed your children for the past five years without them. The investment in labour, which can be prodigious, to build a system to cope with the occasional drought seems less than worthwhile. And even where such systems have been built, who maintains them when they are surviving – even prospering – without them? The systems fall into disrepair.

And then disaster strikes. A year or two of drought, with little or erratic rain, brings few crops in. Weakened by malnutrition, the men and women who farm the land are in no state to start innovating, or to repair the water-spreading devices their fathers once built. After two decades of on-again, off-again drought, despair is almost – the Bishnoi excepted – inevitable. Many then opt for the squalor of the nearest city.

This depressing conclusion, however, is not the end of the story. There are techniques that could be used to grow crops where none now flourishes. More could be developed. Perhaps even more important, there are ways of getting farmers to develop these techniques themselves, to work together to manage their water resources better, and thus to build better, safer lives.

These approaches do not require massive conditional loans from the World Bank; they do not involve vast dams, huge reservoirs, Western "experts" or credence in now discredited theories of economic take-off. But they do depend on grass-roots, local organizations which can stimulate action and demonstrate success.

Spin-off from the Negev

The Israelis have not been slow to realize that the ancient run-off farming techniques discovered in the Negev have considerable significance. They mean, among other things, that Israeli agricultural production could be substantially increased if the Negev were more intensively farmed.

Much research has now been done, mainly at the Jacob Blaustein Institute for Desert Research at the Ben Gurion University of the Negev. Three experimental farms there are devoted to research and development of water harvesting techniques. Two of the farms – the Avdat and Shivta farms – are based on reconstructions of ancient farms first developed by the Nabateans some 2400 to 1500 years ago. The third farm was founded later, in 1976, to demonstrate the applicability of ancient systems to modern needs.

Over the past 30 years this research has succeeded in growing, in the desert, crops such as olives, almonds, apricots, peaches, pistachio, pomegranates and grapes, wheat, barley, sunflowers and a number of fodder crops.

Some of the results have been impressive. For example, one collection cistern, which was probably used originally for watering camels on a caravan route, has now been restored. With a collection capacity of 440 cubic metres, and a catchment area of only 1.2 hectares, the system now waters 300 to 350 sheep annually. Water is directed to the cistern by lines of heaped gravel running down the slope, and by two bunds directing it into the cistern.

The research has, however, not been wholly centred on the Negev; there has been a bold attempt to adapt Negev techniques for developing countries. According to one recent report:

The major goal is to assess biomass production per unit of water, time and land at different production levels from the socio-economic point of view (family, village and district) and from the environmental point of view (mineral limitations, water limitations and no limitations) . . . Special emphasis will be put on optimum production of food, fodder and fuelwood.[3]

The Negev researchers are cooperating with countries such as Botswana, Burkina Faso, Cameroon, Kenya, Mali and Niger in an effort to develop techniques and plant varieties suitable for use in other areas. The Institute also runs three-week training courses on water harvesting for participants from developing countries. Similar research in the United States has concentrated on the water-spreading techniques used by American Indians in ancient Mexico and the south-west of the United States, and has also investigated their potential for other countries.

In a few places, Negev techniques have been directly transported to other regions that also share a winter rainfall. One is Khost plain in the Paktia province of Afghanistan. There run-off is collected in channels from stony and barren hillsides very similar to those in the Negev. The water is led to level fields surrounded by an earth bank, and then passed from one field to the next. Once the soil is wet, a cereal crop is planted and is grown on stored soil moisture and further watering from subsequent rains.

From Israel to the Sahel

The story of how Negev research has been adapted for use in the tropics is more complicated. It begins in the nineteenth century, when French travellers in southern Tunisia first noticed olive trees being grown in what are now called micro-catchments: small, square areas bounded on all four sides by earth walls 20 or 30 centimetres high. Earth is scooped out to form a depression at the lowest point, and a single tree planted there which catches the run-off from the whole area.

Researchers in the Negev have investigated this system, and used it to grow fruit and almond trees in the area. They have compared yields from a range of catchments, varying in both shape and in size, from 16 to 1000 square metres; and further experiments have been carried out in Australia, India, Kenya and North America.

In Kenya, trees have been successfully grown in micro-catchments of 100 square metres in areas that receive less than 400 millimetres of rain a year. In most tropical areas, smaller catchment areas are required than in the Negev, because rains are generally heavier. And provision must be made to channel away excessive run-off when it rains hard. The technique has also been adapted to growing grass on bare slopes by forming "eyebrow"-shaped ridges in which planted grass thrives because it receives run-off from a larger area.

The Negev technique of collecting water, and flowing it from one terraced field to the next has also been tried out – with some success – in Kenya and West Africa.[4] Projects in Baringo and Katui in Kenya have proved the system, though it has not been enthusiastically adopted by local farmers, and it remains to be seen whether they will adopt a technique that requires precision and more maintenance than they are used to.

One of the many problems with schemes of this sort, which can also be found in the Tigray area of Ethiopia and in Turkana in Kenya, are due precisely to the problems identified by Robert Chambers at the beginning of this chapter. Dryland peoples are rarely simple arable farmers. In many places, permanent cultivation on settled sites is not part of the culture, and there is little motivation for maintenance of water harvesting structures that are infrequently visited. A journey back in time over 2000 years to ask the Negev Nabateans about the systems they devised, which combined limited nomadism with careful maintenance of some complicated water collection schemes, might provide an answer. But short of time travel, researchers can currently only speculate.

The Brazilians have borrowed technology not from the Negev but from the Sudanese system of *tera* inundation (see Chapter 9). A government research organization, the Agricultural Research

Centre for the Semi-arid Tropics (CPATSA) in Petrofina, in the semi-arid area of north-east Brazil, has produced designs for the bunds that retain the flood water in the fields, known as *vazantes*, and recommended cropping methods that make maximum use of the available water.

But the most successful of all the Negev-inspired techniques happened when an Oxfam officer visited the experimental work in the Negev in 1979. So impressed was he that he arranged for 200 micro-catchments to be built in seven villages in Burkina Faso, where he was working. Although the original idea was to grow trees for fuelwood in the very deforested area in which Oxfam was working, the farmers soon got other ideas; they began to plant trees for fruit, nuts and fodder. In this, they resembled the Bedouin who, when shown fruit and almond trees successfully grown using micro-catchments, simply said: "But why not plant forage shrubs for our goats and camels?"

The farmers in Burkina Faso also planted millet in their micro-catchments – and some of them, by accident, introduced sorghum. This grew very successfully.

The project then began to evolve rapidly. As it became clear that farmers were interested primarily in grain, attention switched away from trees. A change in Oxfam staff also occurred, and the incomers argued that micro-catchments required too much labour if grains were going to be grown.

Instead, a traditional technique of run-off farming that used to be practised in the Yatenga area of Burkina Faso was adapted. Long lines of stones were placed along the contours on gently sloping ground to slow down run-off and distribute it evenly over a large area. The water was able to seep slowly through the stone banks, and thus make its way through to the next area. Oxfam workers taught the farmers how to mark out a contour using a water level made from a length of clear plastic hose and two sticks.[5]

As the work evolved, Oxfam thought seriously about how to spread the technique. It settled on two-day training courses for farmers which:

- explained why contour barriers worked, using a miniature model and a watering can
- gave practical training on surveying with a water level
- visited places where contour barriers were working
- encouraged farmers to plan their own work.

Peter Wright, who carried out much of the Oxfam work, has been pleased with the results, and with the way farmers took away Oxfam ideas and adapted them to make more complicated structures "without the need for calculations and with a minimum of frustration". He adds: "Run-off from watersheds of up to 10 hectares have been exploited by 'second year' farmers in the Yatenga".[6]

The technique has now spread fast through northern Burkino Faso and on into neighbouring Mali. The technique has also been introduced to Mauritania.

The importance of this project is that it kills two birds with one stone: it increases yields in dry periods; and it controls run-off and prevents erosion when rains are heavy.

This two-pronged attack is the key to the UN Food and Agriculture Organization's recently announced programme to tackle the "conservation and rehabilitation of African land".[7]

FAO rightly argues that in the past efforts to increase yield have led to, or suffered from, rapid soil erosion; on the other hand, attempts to control soil erosion with terracing and other physical measures are rarely successful because the labour involved is huge and the return, if there is any, takes years to materialize. Hard-pressed farmers find little to motivate them in back-breaking schemes that do nothing to benefit them – even if they may, eventually, benefit their children and grandchildren.

Yet there do exist techniques that prevent erosion and increase yield. The Oxfam project in Burkina Faso has demonstrated one. Finding others is a matter of political urgency. It could well be that techniques working on this duel principle could be adopted in many parts of the world's drylands where water scarcity, declining food production and destruction of the environment are proceeding hand in hand.

An almost perfect example comes from nearby Niger. Since 1975 the US voluntary relief agency CARE has been working there to reduce wind erosion from the dry and dusty harmattan wind that sweeps along the millet-growing Majia Valley.

This wind is a serious threat to the precarious water balance of the region. It drys the land, and its crops, and often blows away the lighter, organically rich portions of the soil – leaving behind the stuff of which deserts are quickly made, sand.

CARE has spent $1 million in helping the villagers build 250 kilometres of windbreaks to slow down the wind and protect the fields. Some 3000 hectares of fields are now protected, and millet yields have increased by 15 per cent, bringing a yearly cash increase equivalent, in the early 1980s, to more than $100,000 a year – roughly a ten per cent return on the original investment.

Furthermore, the trees themselves – neem trees from India – can be thinned every few years. When they are, they bring in copious fuelwood and sales of timber that net another $80,000.[8] With such incentive to protect their land, Majia Valley villagers may well be planting windbreaks for a long time to come.

Water Prediction

The results of development efforts often run contrary to common sense. Few casual observers would give a high rating to any scheme that involved Sahel farmers in the latest of modern communications technology. Yet one project in Mali has shown that villagers in 16 villages, equipped with modern radio receivers, can make radical improvements in crop production.

The project depended on a new organization, the Regional Centre for Agrometeorology and Hydrology (AGRHYMET) in Niamey. The centre broadcast information about weather and soil moisture, and provided advice about when to plant, weed and thin crops. The farmers were simply given the radios, and asked to cultivate half their land normally and follow what radioed advice they cared to take on the other half.

The success of this extraordinary idea was such that nearly all the farmers abandoned their traditional techniques as soon as they saw the results on the areas where they followed advice.[9]

Change the Crop, not the Land

Wherever rainfall is limited, crops must be carefully chosen. Where irrigation is used it pays to ask penetrating questions about the relationship between the amount of water used, the area irrigated, the numbers of people employed and the incomes of the farmers involved.

One of those who has is the Indian engineer, V.B. Salunke.[10] He makes a convincing case that sugarcane, the cash crop so widely grown in Indian irrigation schemes, is just about the worst choice of crop in terms of the volume of water needed, the number of people employed and the income produced. If potatoes, for example, were grown instead of sugarcane several things would happen. First, the volume of water used to grow the sugarcane could be used to irrigate an area between five and ten times larger, simply because potatoes are less water-intensive than sugarcane. If this were done, employment would triple and income quadruple.

Growing other crops, such as the Indian staples gram and pearl millet, would have even more impressive results. Pearl millet would increase the irrigated area 30-fold, increase employment eight fold and triple income.

This principle can be extended both to other regions and to non-irrigated areas. Many Africans already grow low water demand crops, such as sorghum and millet – and elsewhere, as in south-west France, farmers are answering a decade of drought by abandoning traditional crops such as maize in favour of these dryland crops. But such has been the neglect of dryland agriculture that, as yet, there are few improved varieties of either these or other dryland crops. Perhaps more importantly, there may well be crops never grown in the drylands that could flourish there.

There is a major research programme into arid zone plants in the Negev. One of the aims is to identify new crops that could be grown in the desert, and to adapt existing ones to desert conditions. "The last fruit of the desert that we domesticated", says Yosef Mizrahi, professor of plant physiology at the Ben Gurion University of the Negev, "was the date palm, and that was over 2000 years ago."[11] Mizrahi is particularly interested in fruit and nuts, and has six major ones under investigation:

white sapote (*Casimiroa edulis*), comes from an evergreen tree, has a large seed surrounded by creamy flesh and is already marketed in Mexico and California;

yehiv (*Cordeauxia edulis*) from the Somalian Desert, provides tasty, easy-to-shell nuts used as a staple by local populations – the leaves also provide forage, and the wood is used for fuelwood and provides a dye;

marula (*Sclerocarya birrea*) a native of southern Africa, related to the mango, and used to make a drink by local people which probably provides 80 per cent of their vitamin C needs;

pitahaya agria (*Stenocereus gummosus*) is the fruit of a cactus – unlike the prickly pear, it loses its spines when ready to eat and has soft edible seeds like those of a fig; texture and taste said to resemble the banana;

mongongo nuts (*Ricinodenron rauteanenii*), the staple diet of the !Kung bushmen, who eat 300 a day, and get 1260 calories and 56 grams of protein from them, the equivalent of nearly half a kilogram of steak;

ber (*Ziziphus mauritania*), an astringent, fibrous fruit coming from a tree that also provides forage – salt tolerant and widely grown in India.

And these are just a few of the fruit and nuts still unknown to nearly all those who live in the world's drylands. In the

water-scarce twenty-first century, arid plant research will need
to become more than a minor backwater for a few specialists.
A major international research effort will be needed to identify
the grains, vegetables, fruit and nuts that can best survive arid
conditions, and mature in the short time for which water can
be made available after the summer rains.

It may be equally important to concentrate on crops that can
be grown under very salty conditions. Such crops could even
be irrigated with salt or brackish water. Dr Dov Pasternak,
of the Boyko Institute for Agriculture and Applied Biology,
also part of the Ben Gurion University, has been cultivating
seawater-irrigated plants for many years. "I am trying to make
people see that salt is beautiful", he claims of the 150 species of
plants he is subjecting to saltwater irrigation.

Research so far has found that a salt bush from Baja, in
California, can be successfully cultivated in salt marshes and
is palatable to both sheep and camels. In Israel, this research
is concentrating on salt-loving fodder crops. At the University
of Delaware in the United States the hunt is on for cereals that
can be irrigated with salt water; at the University of Arizona,
researchers are investigating oil plants that could be irrigated
with salt water.

The water from underground aquifers in desert areas is
often high in salts. It may therefore be of great import that
successful crops of asparagus, broccoli, sorghum, olives, pears
and pomegranates have been grown with saline irrigation.
Limited amounts of salt have even been found to increase
cotton yields, by up to 20 per cent.[12]

Mobilizing the NGOs

Of all the major international action programmes connected
with water and land degradation, the most conspicuous failure
has probably been the UN plan to halt desertification. After
ten years of this multi-million dollar effort, the number of
real successes could be counted on a pair of hands: a few
windbreaks here, a stabilized sand dune there, and a faltering

growth in the number of villages growing their own woodlots was about the sum total of success.

By 1984, seven years after the Action Plan was meant to have been put into effect, only two countries had actually prepared the anti-desertification plans recommended by the UN. Although the United Nations Environment Programme tried to assess progress, the data it obtained from participating countries were so poor that the agency was unable to draw any sensible conclusions other than that desertification was still proceeding apace, and that things had got worse rather than better in most places.

But non-governmental organizations got a pat on the back. According to UNEP:

> . . . NGOs have been the most effective agencies in the campaign against desertification . . . Their high record of success is related to small-scale and local direction of their projects and the requirements for local community participation, as well as their flexibility in operation and their ability to learn from other's mistakes. The dominance of field activities gives these actions an impact out of proportion to the money invested.[13]

Not that UNEP was the first to discover this rather unexceptional fact. The same story had been coming back for more than a decade from projects in Nepal, Peru, Malaysia, Thailand, the Sahel and countless other regions. Where small-scale agencies, which bothered to talk to villagers and farmers, and find out what their needs were and how they might be met, went into action, the results were often better than predicted.

Nowhere is this better illustrated than in Thailand where the International Drinking Water Supply and Sanitation Decade has met with uncharacteristic success. Irene Dankelman and Joan Davidson write:

> By 1990, it is estimated that almost every village in Thailand will be able to provide a minimum of two litres of safe drinking water each day for each person, as well as basic sanitation for every household. This success is due to the local, decentralized approach of working through district and village organizations

with funds allocated directly to them and to appropriate supporting institutions.[14]

There are many reasons for the success of NGOs in the field, and many of them have to do with communication. Local NGOs tend not to walk away from projects that look likely to fail even though they may be short of money and time. Their reputation depends on success, and their success depends on their being able to go back to villagers and farmers, time after time, to sort matters out. Listening is one half of communication – the most important half, it turns out. But for Oxfam's ability to listen in Burkina Faso, it might have persisted in its original policy of trying to introduce tree planting through micro-catchments. Instead it listened to farmers who wanted to grow grain, adapted both its techniques and goals, and came away with a classic case history of successful small-scale development.

The Oxfam story also illustrates another of the keys to forging successful small-scale water projects: organization or – in this instance – the lack of it.

Water use often depends on highly tuned methods of social regulation. Without it, some farmers take too much water, others fail to maintain structures crucial to the whole scheme, some fail to pay their dues and others try to collar too high a percentage of any profits. These problems can apply as much to rainwater collection for domestic purposes as to irrigation development, where they are more common. Many traditional societies have evolved elaborate and equitable methods of social regulation to meet these issues. Others have not – and particularly those where there is no tradition of community involvement with water problems.

One important reason why the Burkina Faso experiment worked was that the technique evolved needed no kind of cooperation. Individual farming families could easily build banks of low stones on their land on their own. They needed neither help from others nor financial assistance (save two days of training and some clear plastic pipe for a water level).

Another example comes from India, in the Kosi region just

south of Nepal. Ground water is plentiful in the region but had to be reached, until 1969, by expensive metal tubewells. Until, that is, a smallholder called Ram Prasad Chaudary Jaiswal invented the bamboo tubewell. This consisted, quite simply, of strips of bamboo bound together by cloth and tar which could be used as a well lining in place of expensive steel pipe. This reduced the cost of sinking and lining a well from 8000 to just 50 rupees at a blow.

Local government officials – though not the Indian government itself – encouraged the development. In the Kosi region alone, the number of bamboo tubewells rose to 1438 in 1970–71, to 19,500 in 1972–73 and to 55,591 by 1977–78.[15] Meanwhile, the large-scale Kosi reservoir project, designed to irrigate 1.4 million hectares, was failing as its distribution canals silted up.

Building on Success

It is now nearly 20 years since a UN agency, UNRISD,[16] cited mutual aid groups as the most important component in projects for the very poor. Over the years their wisdom has been amply demonstrated. And it is a wisdom that applies particularly to water-related projects, where organization is often critical.

The Gram Gourav Pratisthan ("Village Pride Trust") projects near Bombay are a case in point. In this poor farming area, many villagers had given up their farms and gone to live in Bombay. One reason was that they could not afford to irrigate their land because pumps were too expensive for their small plots, many of which were no larger than a fifth of a hectare. The project was designed to build small dams across streams, and allow the collected water to sink through sandy soil to raise water in the wells. This method of water storage, as we have seen, has many advantages since water underground cannot evaporate. But it has then to be pumped and piped to the crops. This is economic only if as many as forty families can share the cost.

The solution was to catalyse farmers into forming their own cooperatives. A local water-users organization, the Pani Panchayat, provided these groups with loans and helped them establish legal title to the water in their wells. It also provided technical help and other assistance.

By 1984 the scheme had grown considerably: there were 52 water management groups, comprising 1725 families, farming a total of 1333 irrigated hectares.[17] This was a self-replicating success, too. Once the first few groups had been formed, and could demonstrate that the problems could be overcome, new groups soon formed without encouragement or stimulation by the parent body.

Small though this project was in scale, it did something that large-scale schemes have consistently failed to do over the decades: stimulate local development, and halt the drift to the cities. It brought back from Bombay many of the families which had fled there to escape rural poverty.

The Mossi Plateau in Burkina Faso differs from the Yatenga area in many ways. One is that its villages have a tradition of forming large cooperative groups to carry out big projects. Since 1977 a novel group called the Six S's (Se Servir de la Saison Sèche en Savanne et au Sahel) has been capitalizing on these groups to speed the development of many water-related projects – small irrigation schemes, reforestation and erosion control. They run schools to teach village leaders new techniques, provide informal forms of finance, and – unusually – stop providing money if it gets badly used.

Their success is simply documented. In just eight years, the movement had grown to include 700 groups in Burkina Faso, 300 in Senegal and 200 in Togo.[18]

A similar success story surrounds the introduction of larger rainwater collection tanks for domestic use in Indonesia. There a local NGO, Yayasan Dian Desa (YDD), has successfully introduced rainwater collection tanks in regions where no traditional roof collection used to be made. And although rainfall is heavy, finding supplies of potable drinking water is very hard at certain times of the year. This was an unusual project in that, though the tanks proved popular and the means of getting villagers to build them were successful, the technology itself has not stood the test of time: bamboo-cement tanks develop leaks if the bamboo is allowed to get wet and rots.

YDD developed the idea of a bamboo-cement tank after discussions with villagers pointed out that ferroconcrete tanks were too expensive. The new tanks resembled the cassava storage bins traditionally used by the villagers, which were made from woven bamboo. The YDD technique is to provide materials and transport, and to train construction supervisors to help the villagers make their own tanks.

And in Thailand the Community Based Appropriate Technology and Development Service has been able to introduce 6500 large bamboo-reinforced concrete rainwater tanks in just four years.[19] One of the keys to this project is that many houses in Thailand are traditional collectors of rainwater from their roofs. But the vessels they could buy for storage were too small for their needs, and they did not have an available technology to construct larger ones economically. The other key is that villagers were persuaded to work together, in groups of 10–15 people, to build the tanks.

Where there exist elaborate forms of social organization to deal with water issues, development is often much easier. The classic example must be Bali, where the most sophisticated form of village level irrigation has existed for centuries. All farmers who take water from the same stream or river are members of an organization called a *sebak*, which meets every 35 days and has its own system of law. This organization plans planting times, distributes water equitably and fines those who cheat. The distribution system is that each 0.35 hectares of land is entitled to a *tektek* of water – the amount of water that can be delivered by a gap four fingers wide which is cut into the trunks of the coconut trees used to deliver the water.

About the only problem with the system is that it depends on traditional stone structures to dam streams and divert water, which are often swept away. Much successful work has gone into improving these basic structures, which have allowed the Balinese to plant their crops at the right time and thus to increase cropping intensity by up to 80 per cent. Writes Paul Harrison:

> In all the small-scale irrigation projects labour-intensive methods are used. Excavation, even of long tunnels, is by pick and shovel; waste is carried away in straw baskets on the head. Masonry is used instead of concrete; and the stones are collected or broken up on or near the site. One Balinese weir was getting the final touches when I saw it – workers were handsetting thousands of tiny pebbles into mortar to provide an attractive finish.[20]

Labour Cost

"Like teams of ants crawling across a hillside" was one visitor's description of the thousands of Ethiopians who took part in that country's ambitious soil conservation programme of the 1970s and 1980s. The programme was supported by several national and international donors and agencies, and it took place in the country's badly eroded central highland plateau where much of the Ethiopian population lives and where much of its grain is grown.

The key to the programme was that the World Food Project supplied farmers with grain and oil in return for each day's work building terraces. The country's 8000 Peasant Associations were used to mobilize and organize the workforce. And astonishing results were achieved. In all, by the early 1980s, 700,000 kilometres of terraces had been built and more than 500 million trees planted.

Massive though this scheme has been, there is much to criticize. First, the cost has been high – far too high for this to be considered any kind of model for other arid areas. Secondly, although there has been participation – in the sense that farmers have done the work, often enthusiastically enough, in return for food – they have been little involved in planning the projects. Nor have they carried on the work once the food for work stopped flowing in. And nor, yet, has the project really done much to increase yields – though the reforestation of the landscape should provide many long-term advantages.

The scheme, in other words, has not generated a self-perpetuating momentum – and, without that, it is unlikely to be successful in the long term.

Tunisia: A Sad Tale

At least the Ethiopian conservation programme got done. Other places have been less lucky. Central Tunisia, for example, is fed by two large wadis, the Zeroud and the Marguellil. They serve 800,000 hectares and half a million people. The supply system used to be intricate, with more than 12 dams, some dating back to the eighteenth century, providing water to those who needed it. The system irrigated 3000 hectares on the Gammouda plain, and many more in other regions, and fed groundwater systems which supplied water to tens of thousands of wells.

There were two plans to develop the region. The first envisaged decentralized management for the Zeroud basin, using available manpower, and consolidating and improving the existing system at low cost. Thirty projects would have been involved which would have irrigated 30,000 hectares, provided 40,000 jobs, raised groundwater levels, slowed down erosion rates and stemmed the rate of rural depopulation. The cost would have been $12 million.[21]

The scheme was ignored in favour of building one massive dam, the Great Sidi Saad Dam, at a cost of $120 million. The scheme provided for the irrigation of only 4000 hectares, of which only 1000 were still in use in 1983. Erosion accelerated, groundwater levels fell and rural depopulation continued. The dam has begun to silt up, thus threatening the supplies of drinking water for the coastal regions, which was the scheme's one major advantage.

The story of the Margellil wadi is even more tragic. In the 1960s, in conjunction with USAID, an integrated management plan was produced which would have developed the 150,000 hectare region. There were to have been 40 hillside lakes, watercourse management, terraces and bunds to slow down erosion, reforestation schemes and agroforestry development to boost the incomes of the 50,000 people who lived in the region.

Work began in the 1960s but was abandoned in the 1970s. In the 1980s a decision was made to focus development plans on yet another huge project.

Enter the Small Dam

History has shown that there are alternatives to the large dam. But recent history has also shown that the development of hydropower does not depend on massive structures, high government (or World Bank) funding, and the displacement of tens of thousands of people.

China has been the leader. Since 1950 China – or, more correctly, the Chinese – have built 6 million small ponds as well as 89,000 minihydropower plants with a total capacity of 6300 megawatts. Each generates less than 50 megawatts, and many are very much smaller. Some produce only a few tens of kilowatts, but still enough to light a village, power a television, and run pumps for irrigation. Most have been the result of local initiative, and have received little or no funding from central government. According to recent figures, some 40 per cent of China's small towns and one-third of its counties receive most of their power from minihydropower units.[22]

The growth in the number of these plants has been astonishing. In 1949, there were about fifty, with a total installed power of some 5 megawatts. By 1977, there were more than 60,000, nearly all of which have been constructed since 1966. One of their many advantages is that they are built from local initiative, and use local labour and expertise. Communities which construct these plants are rightly proud of them. And because the construction process involves those who install them in acquiring expertise, they emerge equipped with the motivation and the knowledge to keep them in running order.

Few of China's minihydropower units break down; and when they do, they are usually quickly repaired.

Pakistan has followed China's example. Even in the United States, the trend now is towards the small dam. By 1980 smallholders there had built an estimated 2.1 million ponds for livestock, irrigation, fish production, recreation and other purposes. Generous tax allowances have also encouraged the development of minihydropower plants; more than 300 megawatts have been installed in recent years.[23]

Japan has given an old technology a yet further twist by resorting to the use of rubber, inflatable dams, filled either with air or water. More than 1000 are now installed, providing irrigation water, flood control and groundwater recharge. One of their major advantages is that they can be deflated, allowing accumulated silt to pass out of the reservoir and flow downstream. The dams are anchored to the riverbed, and can be used to span distances of up to 135 metres. One 40 metre wide dam on the Mekata River is two metres high and is used to generate electricity. The technology is to be exported to both south-east Asia and Europe.[24]

Small dams don't necessarily need new technology. Often, they need social innovation instead. The role of women in water management – like water itself – has been criminally neglected in most development efforts. Yet women are, in effect, the local scale managers of water distribution and collection. Experience suggests that when they take action into their own hands, the results are often spectacular. In Burkina Faso's Yatenga plateau, where water is often in short supply, the women of the village of Saye finally despaired of the lack of village action to build a small dam to ensure the village's water supply in dry periods. They organized a meeting to discuss building a dam, told the men that the women would build a dam themselves if the men would not help, and threatened to return to their parents' villages if nothing was done. They were not prepared, they said, to carry on fetching water over long distances, pounding grain and carrying fuel.

The women won the day, and as a result of their meeting

a dam was finally built. Other nearby villages soon followed suit.[25]

Micro-irrigation

Irrigation efficiency is badly in need of improvement. And salinization must be better controlled if irrigated agriculture is to continue to make an important contribution towards world food security. One answer is the adoption of micro-irrigation, a term confusingly used to describe two different approaches: first, the use of irrigation sprinklers, and of tubes running under the soil which drip water into the root zones of plants; the other is the use of irrigation on a very small scale by farmers and gardners that enables them to raise their own vegetables in areas that are too dry for rainfed agriculture.

More efficient irrigation depends essentially on not flooding a crop but spraying it or dripping water on to it. This form of micro-irrigation is expensive but can double irrigation efficiency – and, therefore, by implication halve salinization and waterlogging problem. By the early 1980s about half a million hectares of irrigated land were being treated with either sprinklers, drip or trickle techniques. While this figure is small compared to the total irrigated area, some countries have adopted the technique in a major way: Israel, for example, uses micro-irrigation on half its irrigated land.[26]

Improved efficiency means that more water is available to irrigate additional land. It has been calculated, for example, that in the Indus Plains increasing irrigation efficiency by 10 per cent would allow an additional 2 million hectares to be irrigated.

Irrigation can be used on gardens as well as field crops. And in many places some very simple technology is all that is needed to provide adequate water to grow such vegetables as tomatoes, aubergines and peppers. Peter Stern[27] quotes an attractive example for a farmer with 10 hectares of land living in an arid area receiving 500 millimetres of rain in a good year and 300 millimetres in a dry year.

An area of 1000 square metres (one-tenth of a hectare) could be set aside to provide an irrigated garden. About 70 per cent of it would be used as a catchment area, and the remainder used to grow crops. The catchment area would lead water down into a 150 cubic metre tank dug in the ground. In a dry year, 210 cubic metres of rain would fall on the catchment area. Allowing for evaporation and losses about 100 cubic metres of water would still remain available for watering the garden – and this would effectively double the 300 millimetre annual rainfall.

Ideas like this have prompted the Common Ground organization to introduce a biointensive system for raising vegetables in more than 100 countries. Depending basically on an organic gardening approach to raising crops, it claims to use only one-quarter as much water and fertilizer as commercial farms, and to raise enough vegetables to provide a complete diet for one person on just one-hundredth of a hectare. Many developing countries would need 30 times this much land to provide a diet for one person. Major research programmes are underway in China, India, Mexico and the Philippines.

Summing up

Rainwater harvesting is about to come of age. Certainly enough experience has been gathered over the past two decades to fill several books with ideas on how to do it. What is still lacking is the year-in, year-out accumulation of experience, under differing field conditions, that could provide a set of choices from which to select when planning new developments.

But if rainwater harvesting is still experimental, at least the need for it is now much more broadly accepted than 20 years ago. Sufficient doubt has accumulated about the wisdom of more big dam projects, and massive schemes to transport water, often uphill, from one region to another.

These ideas are now socially and environmentally un-fashionable. Rainwater harvesting, on the contrary, has an appropriate image about it that meshes well with the gentler ideas of the late 20th century. Because the technique makes use of an untapped resource – precipitation that would otherwise be evaporated before it had a chance to play a useful role in feeding the human population – it looks like getting something for nothing. Making use of such a resource has a certain poetry to it, particularly in a field where the resource itself can never be increased or decreased; unlike food, water cannot be grown to order, even given the right soil and the right fertilizer. But, like food, water can be harvested more efficiently. Doing so is a major priority for the twenty-first century.

Notes

1 FAO, *Soil and Water Conservation in Semi-Arid Areas* (Rome: FAO, 1987, *FAO Soils Bulletin* 57).

2 Chambers, Robert, "Farmer First", *International Agricultural Development*, November/December 1985.

3 Anon, "Desert Agroforestry in Run-off Systems in Israel", *ICRAF Newsletter*, December 1987.

4 Fallon, L. E., *Water Spreading in Turkana* (Nairobi: USAID, 1963).

5 Wright, P., *Report on Run-off Farming and Soil Conservation in Yatenga, op. cit.*

6 Pacey, Arnold and Adrian Cullis, *Rainwater Harvesting: the collection of rainfall and run-off in rural areas, op. cit.*

7 FAO, *The Conservation and Rehabilitation of African Lands* (Rome: FAO, 1990).

8 Timberlake, Lloyd, *Africa in Crisis: the causes, the cures of environmental bankruptcy*, (London: Earthscan, 1985).

9 Timberlake, Lloyd, *Africa in Crisis: the causes, the cures of environmental bankurptcy, op. cit.*

10 Salunke, V. B., *Pani Panchayat, Dividing Line between Poverty and Prosperity* (Pune, Maharashtra: 1983).

11 Quoted in Jeremy Cherfas, "Nuts to the Desert", *New Scientist*, 19 August 1989.

12 "Surprises from Salt Water", *Development Forum*, January-February 1990.

13 Dregne, H. E., *Evaluation of the Implementation of the Plan of Action*

to *Combat Desertification* (Nairobi: UNEP, 1983).

14 Dankelman, Irene and Joan Davidson, "The Invisible Water Managers", in *Women and Environment in the Third World* (London: Earthscan, 1988).

15 Clay, E. J., "The Economics of the Bamboo Tubewell", *Ceres*, 13(3), 43-47, 1980.

16 UNRISD, *The Social and Economic Implications of Large-Scale Introduction of New Varieties of Foodgrain* (Geneva: UNRISD, 1974).

17 Ray, D., *Rainwater Harvesting Project: Socio-economic case studies, volumes 1 and 2* (London: Intermediate Technology Development Group, 1983 and 1984).

18 Timberlake, Lloyd, *Africa in Crisis: the causes, the cures of environmental bankruptcy. op. cit.*

19 Pacey, Arnold and Adrian Cullis, *Rainwater Harvesting: the collection of rainfall and run-off in rural areas. op. cit.*

20 Harrison, Paul, *New Scientist*, 17 November 1977.

21 El Amami, Slaheddine, "Changing Concepts of Water Management in Tunisia", *Impact of Science of Society*, no. 1, 1983, *Managing Our Fresh-Water Resources*.

22 Worldwatch Institute, *State of the World 1987* (New York: Norton, 1987).

23 World Resources Institute, *World Resources 1986 – an assessment of the resource base that supports the global economy* (New York: Basic Books, 1986).

24 "Inflatable Dams Furnish Flexibility", *World Water*, vol. 10, no. 7, 1987.

25 Dankelman, Irene and Joan Davidson, "The Invisible Water Managers". *op. cit.*

26 Meybeck, Michael, Deborah Chapman, and Richard Helmer, *Global Freshwater Quality* (Oxford: Blackwell, for UNEP AND WHO, 1989).

27 Stern, Peter, *Small-scale Irrigation* (London: Intermediate Technology Development Group, 1979).

Chapter 11

Towards World Water Security

*Famine is not about lack of food; it is about lack of
access to food.*

Daniel Nelson, Development Forum
January–February 1990

This book has shown that water ranks alongside food and
energy – if not above them, for both depend on water – as a
key global resource.

Yet water has trickled through the hands of those who
should have addressed it as a global issue. Water issues
have played a backstage role to today's major international
actors: energy, food, forests, fuelwood, and the environment,
for example.

The problem has not been that no attempts have been
made to highlight the issues. First there was the International
Hydrological Decade; then the United Nations Water Conference in Mar del Plata in 1977; this was followed by the
transformation of the IHD into a continuous programme; and,
most recently, there was the International Drinking Water and
Sanitation Decade of the 1980s.

But what, in fact, has been achieved? To paraphrase a recent
BBC documentary, if the thirsty could drink words, there
would be no problem. As it is, there have been more words
than solutions.

International Action . . .

The International Hydrological Decade was coordinated by

UNESCO, starting in 1965. Its objective was to elucidate the workings of the world hydrological cycle. As a research effort, it was a success.

Ten years later the world's water balance was far better understood. According to one estimate, UNESCO's member states managed in that period to make more than 3000 basic studies of water flow in individual catchment areas. To stimulate the effort UNESCO provided education and training for many hydrologists from developing countries. How positive an achievement this was is more controversial. Certainly these training efforts managed to indoctrinate many young hydrologists from developing countries with northern, water-rich concepts.

Nevertheless, without UNESCO's effort much of our present understanding of the immediacy of a world water crisis would not have come to light – other than through the sufferings of the rural populations of the world's drylands. Certainly the limits of the world's water resource would not have been understood as they are now; nor would the regional disparities in water distribution, and the pressing issues to which they give rise.

It was partly UNESCO's stalwart effort that led to the United Nations World Water Conference in 1977 in Mar del Plata, Argentina. It was at this conference that the UNESCO results were first fully reported. Although it was the most significant water resource conference ever held,

> . . . it was poorly funded and poorly managed. It had a budget of about one-twentieth of that provided for the 1979 World Industrial Conference and for a long time had no organizer.[1]

The conference produced 30 recommendations and 10 resolutions. If the conference itself was poorly managed, the follow-up programme was worse. As Malin Falkenmark writes,

> The follow-up has been a major failure. The only problems on which there has been a serious and direct follow-up on

the recommendations of the conference are, first, international drinking water supply and sanitation and, second, international river systems. On the whole, however, the follow-up has been remarkably weak.[2]

The Mar del Plata conference did resolve to supply the world population with an adequate water supply and sanitation within a decade. Hence the UN International Drinking Water Supply and Sanitation Decade (1981–90). The ambitious target was equivalent to providing nearly half a million people with clean water and sanitation every day for a decade.

Although the results are not yet in, all the indications are that the target has been nowhere near met. On the contrary, while many more people undoubtedly now have access to adequate water supplies, thanks to population growth there are also many more people in 1990 without access to adequate water supplies than there were in 1980.[3]

The Mar del Plata Conference also recommended that UNESCO's effort be continued and expanded, and focused more on finding solutions to the world's water resource problems. While it is true that development now features more strongly in this programme, it is questionable whether its enlarged focus has proved as valuable as was UNESCO's earlier concentration on a more specific, scientific issue.

One of the problems is the "e" in UNESCO. Education and training have been principle objectives in UNESCO's hydrology programme. But educating engineers from developing countries to think about water in the same way as water-rich countries does a disservice to the issue (see Chapter 6). UNESCO has still to grasp the real nettle – the need to develop small-scale, local solutions to the issue of water scarcity in the drylands.

But attempts to implement the Mar del Plata Action Plan are still being made. In 1990 a new programme, the Interagency Action Programme on Water and Sustainable Agricultural Development, was formulated by FAO as part of an attempt to implement the Mar del Plata Action Plan for the 1990s. In his introduction to this document, the Secretary General of the United Nations Water Conference, Yahia Abdel Mageed, writes:

> Since its adoption, in 1977, significant progress has been made in the implementation of the Mar del Plata Action Plan. In spite of this progress, however, much remains to be done. The progress in implementation has generally been hampered by a host of complex factors and circumstances, which are often interrelated. Deficiencies in planning and policy frameworks, weak implementation capacities and lack of monitoring and evaluation of performance have generally impeded progress of its implementation. These problems were further compounded by the continuous worsening of the national and international economic environments, the depressed production systems and the occurrence of severe climatic anomalies which have prevailed since the adoption of the Mar del Plata Action Plan . . . . While significant expansion of the area under rainfed and irrigated agriculture has been achieved during the past decade, the productivity response and its sustainability have been constrained by many complex and interrelated factors. Deterioration of irrigation systems and problems of waterlogging and salinization have caused loss of agricultural production. Soil erosion in the upper watersheds, mismanagement and over-exploitation of the natural resources in the drought-prone areas and acute competition for water, have all accelerated the spread of poverty, hunger and famine in the developing nations, particularly on the African continent.[4]

Though these lines may be ponderous (perhaps deliberately so), reading between them is not difficult. Little was done between 1977 and 1990, and the UN system is hoping to make a clean sweep during the 1990s, and try again. Mageed says specifically that the new action programme ". . . aims to rekindle the spirit of Mar del Plata by encouraging a new spirit and commitment for its implementation during the 1990s."

. . . and Inaction

Thus none of these international initiatives succeeded in

treating water as a major but neglected resource. No single international agency has responsibility for water resource issues, and at the national level it too often slips between the fingers of all those who have a peripheral interest in water – ministries of agriculture, of food, of forests, of irrigation, and of natural resources. All have their say but none has control.

Like the proverbial buck, water is too easily passed on to the next player. No one takes responsibility, and no one tackles the key questions – hence, at least in part, the failure of the International Drinking Water Supply and Sanitation Decade.

Yet water resource issues are all inter-connected. It is not possible to separate issues of rainfall and run-off availability from those of industrial and agricultural use; the effects of land use change on water availability must be considered during development plans; and the way water is used and stored inevitably affects downstream flooding, downstream agriculture and downstream fisheries. In the development world, everyone knows this. Yet few plan for it.

To do so, of course, is difficult. Water management is not easy because it must be effected at local level by local people if it is to work. At the same time, water – perhaps more than any other resource – must be subject to overall centralized control and planning. The political leaders of ancient civilizations – like those of today – had to wrestle with these issues. In spite of the fact that they had slave labour to carry out their massive construction plans, water management eventually became too difficult for many of the great river civilizations. There is every indication that, at the end of the 20th century, society is once again facing water issues with which it cannot cope.

The key issue, though, is simple enough: how to provide adequate, reliable water resources for populations living in the world's drier areas. Drinking water is one thing, water for sanitation another – both important. But neither so important as ensuring that farmers and herders in the world's drylands have enough water to raise an adequate crop or herd for four years out of five – and can grow a large enough surplus in those four years to take care of the fifth.

Then, instead of having to fight a daily battle with water

scarcity, farmers and herders could begin to enjoy a feeling of water security. Planning for water security, in place of water scarcity, may be the key.

Food Security and Water Security

There is a potent analogy here with food scarcity. In 1974 the World Food Conference, following the disastrous famines and world food shortages of the early 1970s, created a permanent Committee on World Food Security. And in 1985, the UN Food and Agriculture Organization launched a World Food Security Compact which set out what was required, of governments, non-governmental organizations and individuals to attain food security and eliminate hunger.

Food shortages, it was argued, had multiple causes. And world food shortages were not always the causes of famines. Often there was enough food but it was in the wrong place at the wrong time – as, indeed, proved to be the case in the African and Indian famines of the mid-1980s. And often there were both local and global food gluts, followed by scarcities.

Food security is a concept designed to ensure that food can be made available to those who need it when they need it. It has three main planks:

- increasing production where food is in short supply;
- stabilizing supplies so that countries and individuals can afford to buy the food they need; and
- improving access to food supplies, by boosting rural incomes and reforming land tenure systems, for example.

Thus the idea of simply producing more food to cater for future shortages was considerably broadened.

Food security means ensuring that people can afford to buy food when they need it – this means boosting rural incomes, providing subsidized foods at times, and trying to stabilize world food prices so that countries can afford the food they need in times of shortage.

Finally, food security means putting food aside from good years to provide for lean ones. And this involves ensuring both that there are sufficient global stocks – set at at least 17 per cent of annual world consumption – carried over from one year to the next. It also means providing local food storage warehouses where the glut from one year can be stored for distribution when times are hard.

It is time to re-invent the term water security. Like food security, water security would try to increase water supplies (by techniques such as water harvesting); improve water availability (by techniques such as improved storage); and increase access to water supplies (through improved distribution and price regulation, for example).

Of course, the parallel with food security is not perfect. Water is not traded, as is food; there is no international water market to stabilize; and water cannot (yet) be shipped from one region to another, as can food (to a limited extent).

But some of the main planks of a water security programme could include:

- increasing the availability of water, through techniques such as water spreading, water harvesting and better water management;
- increasing the volumes of water stored, behind dams and via tanks, surface water areas, underground storage and growing biomass, for example, to provide for water needs during dry periods;
- ensuring that development projects include water – as they do fuel, transport and manpower, for example – as one of a checklist of resources that must be specifically provided for;
- encouraging local food storage facilities so that dryland farmers can store their surpluses from wet years and withdraw them in dry years – and thus absolve them from over-elaborate systems designed to grow crops in situations where water supplies are inadequate;
- stabilizing and adjusting water prices so that the poor can afford the water they need, and are not pushed into growing

expensive export crops instead of staple crops to pay for their water;

- creating a Water Aid Agency (with parallels to the World Food Programme) which could supply water aid in the form of financial or technical help to deal with local crises and water famines;
- training an army of professional workers in the techniques of tropical hydology that would be available to provide extension services as and when required in any semi-arid area; and
- providing early warning systems to alert communities to impending water shortages, on both local and regional scales.

Water in the Future

It is not this book's job to invent a political programme that will bring water to the forefront of the world's action agenda. Its job is to illustrate the urgent need to do so. To show that many of the problems that now beset the world's drylands are due to water scarcity is easy enough. To convince planners that the solution of another big dam here, another irrigation scheme there, is no longer adequate is more difficult.

Nor will it do to try to shift responsibility for the present fix on to populations that grow too fast. Population growth is a mixed blessing, putting greater pressures on natural resources but also providing more people to clear land, build houses and grow food. Assuming always that one individual can contribute more to society than an ability simply to feed, clothe and house himself or herself, an increase in absolute numbers does not necessarily pose a threat.

But as has been repeatedly stressed in this book, water is a fixed sum resource: you cannot make more of it and you cannot destroy it. It follows therefore that as human numbers increase, water availability per head must go down.

Yet there are many situations where one man will die of thirst where ten would survive – simply because ten together could dig deep enough to find water, dam a dried-up wadi in time for the next rain, or jointly think up a plan to solve the problem.

As was stressed in Chapter 1, it is human intervention in the water cycle that can increase the volume of water of potential use to society. One place and time to intervene is where it rains, and immediately after it rains, so that precipitation can be put to useful work before it is wasted in evaporation.

The second place to intervene is after the rain joins a river. Without intervention it will then be swept out to sea before it can be used. Conventional planners step in here with the big dam and the new irrigation scheme. Stepping in unconventionally, as the women of Saye did in Burkina Faso (see Chapter 10), perhaps before the run-off reaches the stream, and certainly before the stream reaches the river, is a better bet. Entrusting women, the family-level managers of water politics in almost every society, with the action may also be part of the solution.

The stakes are high. The vulnerability that is induced by water scarcity is the main cause of the famines and crises that occur repeatedly in the dryland areas. Drought and desertification, the conventional scapegoats, are certainly implicated. But they are the symptoms of water scarcity, not its cause.

The prospect of arid land famines spreading, and intensifying, is daunting enough. There is now a real fear that they will soon be swelled by serious international discord caused by water stress. Ethiopia has demonstrated what happens when civil strife and hunger are combined. Water scarcity threatens to produce the even more explosive combination of international conflict coupled to arid land famine.

On the other hand, local solutions to local shortages, providing water security for the 20 per cent of the world population that lives in the drylands, might yet avert this awful prospect. It would be comforting to be able to turn back the pages of history, and find examples of how societies used to manage their affairs on limited water supplies. Such examples exist. But history shows more clearly that many societies failed to solve the problems of irrigation and water control. And several perished as a result.

For those caught up in the famines of the past two decades, history has already repeated itself. When water scarcity forces millons to flee their homes in the Sahel, it is time, surely, to take stock; when millions of lives are put daily at risk from water-borne disease, it is time to ask searching questions; and when the richest – and wettest – countries begin to squabble over who owns the rights to which source of water, it is time to invent new political solutions.

The Koran reminds us that water is the source of all life. The experience of the past decades is a grim reminder that the converse is also true. Managing water better – locally, nationally and internationally – is now a matter of life and death.

Notes

1 McDonald, Adrian T. and David Kay, *Water Resources: issues and strategies* (London: Longman Scientific and Technical, 1988).
2 Falkenmark, Malin, "Global Water Issues Confronting Humanity", *Journal of Peace Research*, vol 27, no 2, 1990, pp 177–190.
3 In 1990 there were an estimated 243 million urban and 377 million rural dwellers without reasonable access to safe water supplies. Ten years previously the respective figures were 213 and 292 million. See *Levels of safe water and sanitation services 1980 and 1990, all developing countries*, UNDP Information Paper, 1990.
4 FAO, *An Interagency Programme on Water and Agricultural Development* (Rome, FAO, 1990).

Appendix I

The Swedish Red Cross and Linköping University
Expert meeting on vulnerability generated by water scarcity in
semi-arid regions.

Vadstena,Sweden, 13–16 February 1989

Summary Report

The meeting which was convened by the Swedish Red Cross
and the Department of Water and Environmental Studies,
Linköping University, brought together people and perspectives that seldom meet within the framework of a conference.
Macro and micro perspectives as well as desk and field out-
looks were presented by researchers in hydrology, agriculture,
forestry and social sciences and from international governmental and non-governmental organizations engaged in relief
and development. Africa, Asia, Europe and America were
represented.

The meeting took as its point of departure a document by
Falkenmark, Lundqvist and Widstrand at Linköping University
that focuses on water availability as a fundamental constraint
to development in semi-arid Africa. The document states that –
given the current population increase at an unprecedented rate
– it is possible to identify regions in Africa that may face water
shortage of extreme severity within a few decades.

In contrast to water's significant importance in socio-
economic development and in the environment, it has invariably
been interpreted in an oversimplified manner in connection

with development efforts. Focus has mainly been on technical supply aspects and on visible water only.

It was concluded that solutions to problems related to water shortage are to be found in natural resource conservation and management that duly consider the complexity, diversity and risk-proneness that characterize the semi-arid environments. Development interventions that do not accept these pre-conditions will not generate a sustainable development. However, when they are accepted, experience has shown that techniques for soil, water and nutrient conservation and concentration will enable biomass production at a much higher level than previously thought possible in the semi-arid regions. In many cases the technical solutions are not new. They have been part of a traditional knowledge that has been disregarded and ignored by conventional development thinking. Appropriate research efforts are largely non-existent, particularly when compared to what has been investigated in promoting agriculture in so-called high-potential environments. Consequently, there is enormous scope for research and interventions aiming at sustained biomass production in semi-arid lands.

It was particularly noted that through appropriate measures, physical rehabilitation efforts on barren land may produce results already in the short term.

In order to realize the potential of the drylands, their resources have to be under the control of local people. Without security of rights and tenure, they will not invest their time and labour in the rehabilitation and development of drylands. Thus, a call for natural resource management is also a call for empowerment and democratization. Particularly urgent seems to be a rapid response to the already existing strong demand among women for child spacing, now reflected in abortions on a massive scale. Another important measure is to develop innovative ways for water resources assessment to be used as a basis for socio-economic development and environmental management. This is a crucial element in long-term strategy.

For any agency, be it governmental or non-governmental, interventions to promote self-reliant development must be based on community problem identification and aim at partnership between community and agency in decision-making and responsibilities. Structured and systematic ways of involving local communities in fruitful dialogue are available, but little appreciated by government agencies. Exchange visits between communities have a great potential in enabling transfer of successful innovations.

It is obvious that conflict at different levels and of different insensities is a major obstacle to any effort towards sustainable development. Most success stories have occurred under relatively harmonious social conditions or after resolution of conflicts. Outside interventions must therefore involve a preparedness for conflict resolution and consensus building.

It is clear that there is a specific role for NGOs to act as brokers and advocates towards governments and donor organizations as regards secure rights and tenure, conflict resolution and new approaches to empower local communities. Measures for integrated soil-water-nutrient conservation and management have to be included in national and international conservation strategies. For relief agencies like the Red Cross there is an obvious need to put drought relief in a development perspective in order to couple early response and security against destitution with long-term rehabilitation and development.

The meeting's participants identified a number of issues that need to be addressed specifically in a process that goes beyond the Vadstena meeting. Among these is the need to create awareness among donors and macro-level policy and decision makers along the lines discussed here.

The meeting's hosts – Swedish Red Cross and the Department of Water and Environmental Studies, Linköping University – declared their intention to establish a core group that will assume responsibility for the continuation of the process started in Vadstena. The core group will also include representatives from the Swedish University of Agriculture Sciences. It is furthermore expected that some of the meeting's participants will function as corresponding members of the core group.

Appendix II

Sustainable Development and Water

Statement on the WCED report
Our Common Future

This statement emerged from discussions during a Special Session on Water Strategies for the twenty-first Century at the IWRA IV World Congress on Water Resources in Ottawa 30 May–3 June 1988. Due to the fundamental importance of Water for Our Common Future, this short statement is being forwarded to the Oslo conference on Sustainable Development 9–10 July 1988. A more complete paper, stressing the urgency of such water-related strategies that are necessary to make sustainable development possible is under preparation, and will be published and given world wide distribution in June 1989.

1. The presence of water distinguishes the Planet Earth from other planets. Biomass production is driven by water and most of man's activities are water-dependent. Water is circulated in and provided from a global system, the hydrological cycle. The type of natural disaster affecting the largest numbers of people are droughts and floods, both water-related. Water flows form fundamental components of ecosystems on all scales. Water, in view of its great versatility as a substance, forms an extremely complex part of both natural systems, societal systems and every-day life.

2. All these facts add up to the basic fact that *human life is constrained by the limits posed by the global water cycle, and the natural laws governing that circulation system.* This simple fact is illustrated by an increasing scale of environmental pollution, generated by water's mobility and chemical activeness. Water is a unique solvent, always on the move in the visible as well as the invisible landscape. The same fact is illustrated by the water-scarcity driven problems in the arid and semi-arid tropics. In fact, water scarcity is expected to develop into a first-rate issue within a few years time only. The Ottawa World Water Congress even concluded that in the 1990s water will replace oil as the major crisis-generating issue on a global scale.

3. In view of what has just been stated, WCED in its report *Our Common Future*, although aware of the fact that water is fundamental for soil productivity, and that water may be subject to resource limits, tends to severely underestimate water-related problems involved. The report is indeed strongly misleading in that respect. In spite of the evident ambitions to cover the specific problems of each Third World region, the Commission *pays no attention to the galloping and multidimensional water scarcity now developing in Africa.*

4. The fact that so many developing countries are situated in an arid and semi-arid climate is thought-provoking in itself. A report discussing a sustainable world development without reference to the specific conditions of these arid and semi-arid regions indeed lacks in credibility. Had the Commission been more aware of the implications of water-related problems, particularly in the arid and semi-arid tropics, it would probably have concluded that *water – being as complex as energy – would have deserved a sub-chapter of its own in the book.* It is in fact remarkable that there are sub-chapters on the oceans, space, and Antarctica but not one on fresh water, the blood and lymph of the geophysiological system that we call the biosphere.

5. Such a chapter would have brought up the many parallel basic functions of water and the many links produced,

creating dilemmas for human activity and global development: the hydrological link between the water consumed for plant production and the water remaining for feeding terrestrial water systems in aquifers and rivers, ie. the water available for human use; the chemical link between moving subsurface water and the quality that it attains; the mobility link between land use activities on one hand, and river response, flow, variability and quality of both groundwater aquifers, rivers and lakes, estuaries and coastal waters on the other.

6. Although there are some scattered references to water scarcity related problems, the *Report as regards water resources is heavily biased towards present thinking in the temperate zone.* The broad geographical composition of the Commission and the substantial input from world-wide hearings were evidently not enough to counter this intellectual inertia.

7. The absence in the WCED Report of a credible discussion of sustainable development as seen from a water perspective caused great concern, even dismay, at the World Water Congress. The Congress therefore urged the Committee on Water Strategies for the twenty-first Century to produce an IWRA paper on sustainable development as seen from a water perspective. That paper will analyse the complex involvement of water in relation to sustainable development, the various criteria and components involved, and main processes and activities that are threatening the sustainability. The report will discuss key concepts such as limitations posed by the environment, sustainable yield, systemic effects, etc. Given the fact that life is based on myriads of water flows – through every single plant, every animal, every human body – and that it is globally circulated in the water cycle with which man continuously interacts by land use and by societal activities, *sustainable development must be a question of sustainable interaction between human society and the water cycle, including all the ecosystems fed by that cycle.*

8. The fact that water is subject to the natural laws controlling the hydrological cycle has distinct consequences for the

degrees of freedom under arid and semi-arid conditions. The effects of such limitations in the hunger crescent in sub-Saharan Africa can be seen in the breakdown of the environmental fabric of the area, and in population dislocations, driven by the spread of a feeling of insecurity among the zone societies. The area is involved in a population-driven risk spiral, composed of several types of water scarcity, added on the top of each other and released during intermittent drought years.

9. *Fundamental strategy changes are needed to address the massive sustainability problems in the realm of water.* Institutional systems have to be resource-oriented and integrated, rather than use-oriented and sectorized. Changes in water attitudes are crucial to avoid the present biases and make sustainable development possible. The legal framework has to take water as a major element in a comprehensive land use law.

For the Committee on Water Strategies for the twenty-first Century

28 June 1989

Index

Water: The International Crisis